Growing Food & Flowers in Containers

Growing FOOD & FLOWERS *in Containers*

BY TED FLANAGAN
Associate Professor of Plant & Soil Science
University of Vermont

Illustrations by Robert Vogel

GARDEN WAY PUBLISHING CO., CHARLOTTE, VERMONT

Standard Book Number: 0-88266-015-2
Library of Congress Catalog Card Number: 73-89124

GARDEN WAY PUBLISHING CO., CHARLOTTE, VERMONT 05445

Designed by Linda Dean Paradee
Printed in the United States
Second printing, July 1976

Contents

Introduction

The many wonderful advantages of growing plants in containers have not yet been fully realized. While grandmother grew geraniums in pots in the house, kept a green fern in a wicker tub in the foyer, and started her tomatoes in tin cans on the kitchen windowsill, today there is a whole new spectrum of methods and advantages available.

Using today's modern containers and easy-to-follow methods you can extend the pleasures of summer flowers and fruit, growing them indoors as well as out, even year-long, right inside the house. You can start your own flower and vegetable seedlings for use later in the garden or on the patio, even to take with you to a vacation home. Container gardening makes available the job of gardening to the handicapped, the convalescent, and offers advantages to those teaching youngsters about gardening and plant life. Growing plants the container way maximizes the benefits of flower, foliage and fruit while minimizing the effort involved.

The key to container gardening is that you can grow all sorts of plants quite successfully, each in its own portable garden. All plants need certain amounts of light, air, water and plant food nutrients in order to prosper, and these you can provide easily the container gardening way.

Just what to do and when, depends a great deal on whether you're starting out in the spring or fall— or perhaps you wish to start indoors during the cold of winter. So the following sections are arranged for convenience in the sequence of following the progression of the seasons.

The concept of container gardening is not new, as you will note in Chapter I. What is new, however, is the variety of values one can gain by these methods. If it's almost spring, you can get a headstart on outdoor gardening by planting seeds in containers so the young plants are ready to be transplanted in a garden out of doors—or in proper containers—when the season is ready.

Chapter II gives all the details necessary to start plants indoors for later use elsewhere, outdoors or indoors, for the garden, patio, house, balcony, porch, schoolroom or summer camp. Simple but precise directions are given for starting plants from seed, seedlings, transplants, slips, cuttings, grafts, and other methods of growing your own plants from scratch. This not only gets a headstart on spring but allows you to grow special flower or vegetable varieties that you can't get elsewhere.

When summer is over but you wish to have it last a bit longer, the ways to extend its benefits are detailed in Chapter III. So that the color, fragrance and flavor of flowers, fruits and herbs are not lost to an early frost, you move prized plants to a more welcome "climate."

House plants usually imply potted geraniums, ferns and the like, but today, due to the versatility of container gardening, many more decorative and useful plants easily can be grown year-round indoors. The hows and when of how-to-do-it are given in Chapter IV. And too, in many localities, it is quite possible and highly desirable to move your "potted" plants from house to patio, porch or balcony, then back inside if the weather becomes inclement, or, for example, if you wish to have a spectacular show of colorful blossoms inside. Specifics on how to put plants where you want them and when it's best to do it (with emphasis and details on container mobility) also are located in Chapter V.

Chapter VI spells out which plants and varieties are best adapted to the several different types of container gardening, and for different uses, such as plants for height accent or horizontal emphasis, plants selected for their attractive shape or foliage and for colorful flowers or fruit. Also the vegetables and herbs are given, listing those best suited when grown in containers for food and flavor, and which plants do best around and in pools and fountains, those indoor-outdoor plants that do equally well in containers on the patio or in the living room, as well as those that can be selected for a useful and colorful existence solely within the house. Noted here, too, are those plants best adapted to miniature container garden arrangements, bottle gardens, terrariums and for bonsai culture.

In addition to the traditional pots, tubs and tin cans, almost any sort of container that will hold sufficient soil will do, and a great diversity of types and shapes of containers is available today. Some you will select for their aesthetic qualities—because they're colorful or of unusual shape; others for special functions such as starting seedlings, or to fit on balcony railings and other special locations.

The selection and care of containers is the subject of Chapter VII. Their function, qualities, and maintenance are simply but fully explained. Especially important, since larger containers may weigh 50 to 100 pounds when filled with soil, are the notes on how to increase their mobility—such as with wheels and casters. Here, too, are hints on the selection of special containers of glass or other material for terrarium and bottle gardens.

What's In The Pot, Chapter VIII, includes hints on special seed-starting and light-weight container soil mixes, how to prepare and care for the soil that goes into the container to achieve a maximum of benefits to the plant with a minimum of effort on your part. Here, too, are details on how to pot and re-pot; how to trim and shape root-bound plants; whether or not to provide rest periods—to allow plants to recuperate to prepare for flowering. Hints about pest control are also presented in this chapter.

There are several special applications of container gardening that provide unique advantages not otherwise available for those who can't garden conveniently, for those confined to wheelchair or to bed. Special note is made in Chapter IX of how to arrange containers for such accessibility. Pots or larger containers holding a whole miniature garden, bring delight to handicapped gardeners or those who otherwise can't get outside to garden in the conventional manner. And a micro-miniature garden of small plants inside its own see-through glass bottle often is a great joy to a person unable to garden actively at all.

Building a preplanned and prepackaged vegetable or flower garden to take along on vacation to your summer camp or cottage is easy with containerization. How to prepare an "instant vacation garden" with plants preferred is programmed in simple steps in Chapter X. In this chapter too, are important suggestions on how to care for plants left at home while you are away. Details of several "automatic watering" methods are given.

Chapter XI outlines how container-grown plants may be useful in the classroom to assist in the instruction of plant life, botany and ecology. Included are simplified procedures for the summer camp counselors who'd like to have some garden plants "ready to go" for their youthful campers.

Container gardening also embraces the art of miniaturization. Success in bonsai, terrariums or gardens-in-a-bottle, depends on exact control of the plant environment. Chapter XII offers suggestions on how to provide this. The special modifications used in bonsai culture are left to the more advanced gardener, but references to excellent sources of information are listed in the Appendix. Gardens-in-glass, a special adaptation of container gardening, are covered in detail in this chapter, with step-by-step instructions on how to make a bottle garden. Their success depends on establishing and maintaining the necessary plant growth conditions from the start. This is done easily in closed containers, each encapsulating its own self- regulating plant life system. The availability of all sorts of glass containers broadens this type of gardening, which can provide, as well, decorative and useful gifts.

At the very back of the book, last but not lease, are listed places where to seek advice, purchase seeds, plants, containers and other supplies. There is a descriptive list of the plants most useful to successful container gardening.

So whether you want flowers, fruits or vegetables, and whether you prefer to grow them conventionally or organically, the choice is yours. How to do it the container gardening way, easily and successfully, is fully explained here. There's no time like the present to get started!

Growing Food & Flowers in Containers

CHAPTER ONE

Whys and Wherefores of Container Gardening

Gardening in containers allows you to grow plants where you want them and when you want them. Very simply, it involves raising plants in pots, boxes, or any sort of receptacle that holds the soil in which the plant is rooted.

Most plants are grown in an outdoor garden or flower bed, but since this miniature "garden" in which your plant will grow is contained in a receptacle, it follows that it can be moved about, indoors or out, into the house from outdoors or vice versa. Hence, the versatility of this technique.

A containerized plant is sort of a self-contained eco-system—you providing the proper environment by moving the plant, container and all, to that location most conducive to its welfare. Don't neglect the fact either that, for short periods of time, you may want to place the plant where it will benefit you the most—indoors for show, for example. For the ultimate goals of container gardening is to provide benefit to yourself—to have flowers, fruits and vegetables when and where you so desire.

Container gardening was favored many centuries ago. From then until now Oriental gardeners have specialized in growing miniature citrus and forest trees in tiny pots or bonsai arrangements. The ancient Romans coaxed roses into bloom in tiny greenhouses of mica, and they nurtured pots of herbs through the winters in their homes.

The Egyptian Pharaohs thought so highly of container-grown fruit trees that they were depicted on the walls of their tombs, but the hanging gardens of Babylon undoubtedly were the greatest display of container gardening.

Potted herbs and fruits played important roles in many a medieval monastery. For practical purposes, plants were overwintered indoors in pots and tubs so to be available for next spring's gardens. But they also served as a winter-long source of herbs and seasonings, as well as plant materials for religious rites.

Louis the XIVth, not to be outdone, ordered his gardener to produce ripe cherries and figs for the royal table in winters. This M. LaQuintinya did, by moving trees in boxes to heated greenhouses.

The landed gentry of England during the last two centuries brought indoor gardening to its zenith with their greenhouses, atriums and orangeries, in which they maintained both useful but mostly exotic plant species. These included citrus fruit (hence the special name orangeries), where they were housed in the off-season. But history also notes that it was not necessary to have special glassed-in edifices to grow plants in containers indoors or out. The common people in England always overwintered their favorite perennial herbs in windowsill pots, not to mention the few flowers that were housed indoors overwinter. When spring finally came they could be placed back once again in the dooryard garden.

Since container gardening is simply growing a plant where and when you want it, one can surmise that over the centuries just about everybody who gardened—and many who didn't—kept plants in the house during forbidding weather. Colonial records tell of favorite old-world herbs surviving the ocean voyage to America. With the conditions aboard sailing vessels of that day, water for plants must have been very scarce, but some survived the voyage, sealed away in Wardian cases—those historic forerunners of our modern bottle gardens!

To bring container gardening history to date, one needs only to recall the Victorian parlors, the sunny farmhouse kitchen windowsills, the terra cotta pots with their collections of potted ferns, geraniums and hen-and-chickens. Remember also, as already noted, the tomato plants in the tomato cans on the windowsill, stretching out toward the spring sunshine. No, container gardening is not new, but today, because of our new knowledge, almost any plant can be grown where you want it, indoors or out—even outdoors some of the time, then moved indoors for spectacular display or to avoid inclement weather.

The versatility and adaptability of container gardening is what makes it a popular and useful way to grow plants. Just about any plant can be grown this way, at least for a while. And just about any sort of recepticle will do, though many sorts of handsome containers now are available.

Of course some decisions must be made. What kind of plant or plants you wish to grow depends on your objectives: what you wish to grow them for—for color, fragrance, fruit, herbs for the kitchen, greens or tomatoes for salads.

Your choice of container, too, will depend in part on appearance, whether it enhances the situation. Convenience, too, plays a part. This depends, again, on what you want to grow and where. A miniature orange tree for the patio might look best in a redwood tub while chives on the kitchen windowsill would do well in any reasonable pot.

Container gardening is adaptable to a number of situations. Much of the historical use of this technique had to do with overwintering prized garden plants and providing bits of color and taste during the otherwise drab winter seasons. The adaptability of modern container gardening is that you can extend your gardening benefits throughout the year, and can put your prized plants where they will contribute most to your enjoyment, be they flowers, fruits or vegetables. And since you will be growing a plant in a container and moving it about, you can take advantage of the climate or environment that benefits both the plant and yourself.

You won't need a green thumb. How to grow a plant is fully explained later on. For it to prosper it needs light, air, water, plant food and a benevolent temperature. These conditions, while they may vary from plant to plant depending on what you grow and where, can be met easily.

The soil in the container where your plant is rooted not only "holds the plant up" but is the plant's source for water and plant food. You simply supply the soil with sufficient water and plant food or fertilizer—which in most cases is very infrequently. Most failures in container gardening result from over-frequent watering and more-than-solicitous feeding.

The other needs of the plant—light, air and temperature—are met by portability of the container. By simply changing its position or location you can provide optimum conditions for its well-being. This is one of the real reasons for growing plants in containers. Not only can you place them where they will benefit you most, but you can move them about to their own benefit. When fruiting or flowering, they can be placed in a special setting to enhance your pleasure. And when *their* needs dominate, they can be moved indoors or outdoors so they will better prosper. All that is required is that the plants and their containers be adaptable to the situation.

No matter where you live, city or country, apartment or house, container gardening can provide a whole array of advantages—from starting your own plants indoors ahead of time, to bringing fall-blooming plants inside to stretch out their summer benefits. With some additional lighting, and the proper containers, both simple to provide, you can start seedlings, root cuttings, have flowers and fruiting plants, exotic foliage plants, climbing vines, tubbed fruit trees, fancy vegetables, even grow a salad garden, indoors!

CHAPTER TWO

Getting a Headstart on Spring

Down through the ages, man always has heralded the arrival of spring. And where dreary winters were long and most inclement, he often tried to hasten the arrival of the first spring blossoms.

Again history records the use, outdoors and in, of jars, transparent cloches, akin to the modern "Garden Aider," to force plants into leaf and bloom.

And who does not recall grandmother's jar of pussy-willow buds or fruit branches, not to mention the dish of narcissus bulbs in their bed of well-washed pebbles, and the young vegetable and flower seedlings in their pots on the windowsill. Plants in containers long have been a way to start a garden and to bring home the joys of spring ahead of time. Long ago an English gardener described this as "advancing growth by simulating conditions not present in the open." In a way he was forecasting the versatility and adaptability of modern container gardening.

Most gardeners, no matter where they are located, would like to have flowers bloom and vegetables to eat as soon as possible. In the cooler regions where frost-free growing seasons are short, this means *sooner* than nature provides. But by means of today's ability to grow plants indoors it is now quite possible for you to grow plants to full maturity where otherwise they would be harvested by Jack Frost.

A common practice is to start your own seedlings and then transplant them outdoors to complete their growth to flower and fruit well within the confines of a short outdoor growing season. This applies particularly to such vegetables as tomatoes, peppers, eggplant, cabbage and its cousins: cauliflower, broccoli, and Brussels sprouts.

Of course, no matter what your season, you can bring your flower garden into bloom weeks ahead of time by starting annuals, biennials—even some perennials—from seed, ahead of time indoors. Such long day-annuals as coreopsis, phlox, rudbeckia, scabiosa, and snapdragons respond well to this technique. The same is true of the short-day annuals, especially cockscomb, marigolds and zinnias, as well as perennial English daisies, foxglove, sweet william and hollyhocks.

Another reason for starting plants indoors is to grow for transplants the varieties and kinds of flowers and vegetables that you don't usually find in local nurseries or greenhouses. Most retail plant sources carry only the most common sorts, and if you wish to have a particular kind of vegetable or variety of flower, chances are you'll have to start them yourself from seed. Your indoor container gardening setup allows you to do this easily.

Last but not least is the fact that pennies for seed means savings of dollars for plants. Even with the most expensive of indoor supplemental lighting, its cost, when spread over a number of seasons, and considering its use for growing plants indoors all year long, is not exorbitant. Careful accounting of the costs of all materials and equipment used to start your own seedlings—the trays, pots, soil, seed, and their share of light and temperature aides—prove that it can be done most economically.

No matter what your reason, it is often most advantageous (and a lot of fun) to start your own garden plants from seed. Using the lights and some of the other container gardening materials you have at hand, you can grow enough plants to satisfy your outdoor garden's needs, have extras to give to friends, surely save some money and definitely get a headstart on spring.

Starting the Seeds

First of all the seed: I'm assuming, of course, that you already have decided what you want to grow: vegetables, flowers or fruit. And you may already have the seed in hand, having selected the kinds and varieties desired. But if you haven't, look through the catalogs and seed racks, selecting only those varieties that will do well in your particular climate.

Where possible, choose new hybrid varieties and those noted to be most disease resistant. Why not benefit from the experience of the plant breeders? Don't overlook, particularly where your outdoor gardening space is limited, the many new midget and dwarf varieties now available.

Next, how to go about it: All seeds are tiny, dormant plants in a state of hibernation, simply waiting for the proper growing conditions to sprout up as young seedlings. These encapsulated, embryonic plants carry with them their initial source of plant food for energy and growth. All that's needed to get them to germinate, to sprout, is a happy combination of light, water and warmth.

Since there is an almost infinite variety of vegetables, flowers and fruits, you might expect a great diversity in their proper germination or seed-sprouting conditions. Luckily, however, most seeds will germinate under the conditions of light, water and warmth very easily supplied in the home.

Many varieties will germinate equally well in the light or in the dark. This is why they come up when planted and covered with soil in the garden or seed flat. Others germinate only in the dark, yet some definitely need to be exposed to light to sprout successfully. Where specific light requirements for germination are necessary, this information is usually noted in the seed catalog or on the packet. Varieties with special light requirements for germination are noted in the Appendix.

Where light is called for, the light from one or two low-wattage (40 W to 60 W) incandescent bulbs usually is sufficient. The light from cool-white fluorescent lamps, fluorescent plus incandescent, or the special plant-growing fluorescent bulbs will do as well. Natural light from a window also will suffice for seed-sprouting, provided that the duration of light available is sufficient and that the direct rays of the sun don't overheat the germinating seeds.

Where light is required for germination, artificial light probably is best, because you easily can regulate the brightness and the length of the lighting period. And you avoid overheating by keeping the light bulbs at a distance, or by using a small fan to circulate the air between the sprouting seeds and the lamps.

Moisture, vital to all seed germination, is provided by the soil medium and the containers in which the seed is sown. This soil medium usually is sterile vermiculite, perlite, sphagnum moss or some happy mixture of one or more of these. Some gardeners use sand, but since this dries out very quickly, and adds unnecessary weight, it should be avoided. The inclusion of soil in the seed-sprouting medium is not necessary. The plant foods or nutrients in the soil are not needed until the seedlings have exhausted their own seed-contained supply. And soil frequently contains harmful microorganisms causing that most disappointing seedling disease called "damping off."

While vermiculite, perlite and sphagnum moss essentially are sterile—that is, free of disastrous disease organisms—soil generally is not, and if included in the seed-starting medium, must be sterilized (see page 15). To be on the safe side your containers should be completely clean and sterile. Boiling water will do the trick. Where new containers are not available (and they are to be preferred) old pots or flats can be soaked, then scrubbed, in cold water to remove soil and green algae stains. Then use a boiling water rinse.

The best container to use is one of the small, standard flats or pans available at garden supply centers or from seed companies. These usually are made of plastic, although some disposable ones are of sturdy papier-maché. Choose those with drain holes in the bottom. In a pinch, milk cartons with one side removed will do, but be sure you punch a number of drain holes in the bottom. The easy way to do this is to cut a few small v-shaped notches along each side with a sharp knife. Deep foil frozen-food pans or foil baking tins will do too.

Notched milk cartons will do in a pinch.

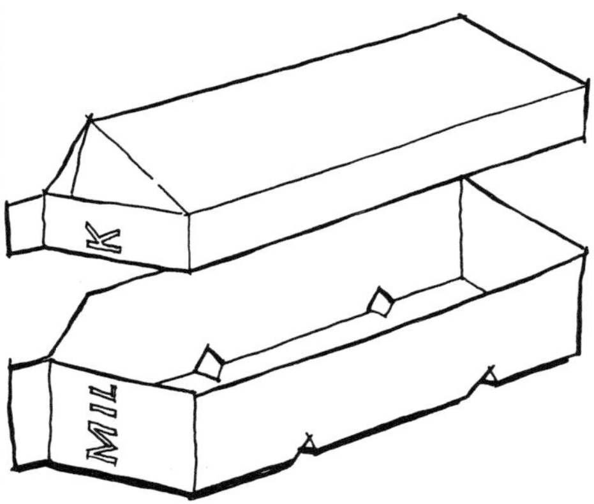

Punch holes for drainage. The best sizes to use are those at least two inches deep and rectangular in shape.

Round, unglazed clay pots also are useful for starting seeds. The best are wider than deep. These are called standard fern pots or 3/4 pots, because they are three-quarters as high as wide. Select those at least three inches deep. But rectangular-shaped containers are preferable for ease of planting the seeds in straight rows.

Large drainage holes in containers can be covered with a layer of newspaper or plain cardboard to prevent the "soil" medium from leaking out. This is usually not necessary if the coarse or insulating grade of vermiculite is used, at least in the bottom of such containers.

To prepare the containers for sowing, first line the bottom with moistened coarse vermiculite, coarse perlite or a mixture of both. Then fill the container to overflowing (it's a good idea to set the container on newspaper to facilitate later cleanup) with finer vermiculite or vermiculite-perlite mixture, or one of the combinations noted above, with or without sphagnum moss. I have found moistened vermiculite alone as good a germinating medium as any.

Then, with a flat stick or ruler, scrape off the excess vermiculite. Next, with a small rectangular board slightly smaller in dimensions than the length and width of your seed flat, press the vermiculite down firmly. This will leave the surface of medium about a half inch below the rim of the container. Lacking a proper flat board, a child's wooden block (the flat kind) will do; or even a small paperback book wrapped in aluminum foil. For pots, a round saucer would do, or the bottom side of another pot of similar dimensions. However, don't use the palm of your hand, as this will result in uneven compression and uneven water provision to the seedlings later on.

It is at this point, or perhaps just after filling, that you may wish to perform an additional sterilization operation by pouring boiling water—carefully—into the container of vermiculite, until it runs out the bottom drain holes. Repressing of the surface may be necessary, and of course the container must be cooled before seeding.

Now to seed: to keep the rows straight (and

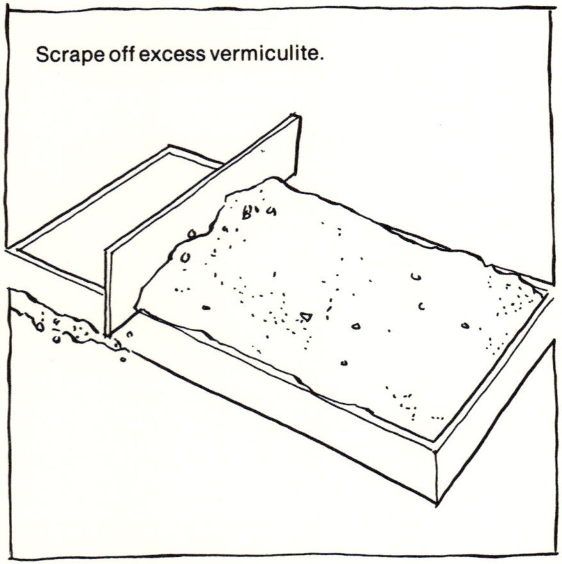

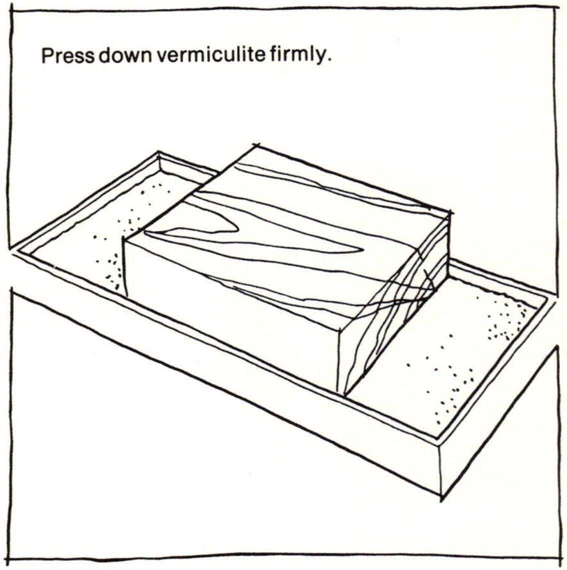

usually you will wish to plant more than one row per container, perhaps of different varieties), small grooves or depressions may be made. These seed furrows or drills also enable the seed to be more easily covered with the medium where such is required.

The easiest way to do this is to make an indentation with the corner edge of the board or by pressing with the fingers a pencil, laid flat, into the surface of the vermiculite. Follow the directions for planting depth found on the seed packet. Larger seeds such as marigold, zinnia or tomato may benefit from a light covering of vermiculite. Some gardeners prefer to cover the seeds with a light topdressing of screened sphagnum. I feel that fine vermiculite does very well where covering is required.

After sowing, firm the medium again with the flat board. This will bring the tiny seeds into intimate contact with the vermiculite, help-

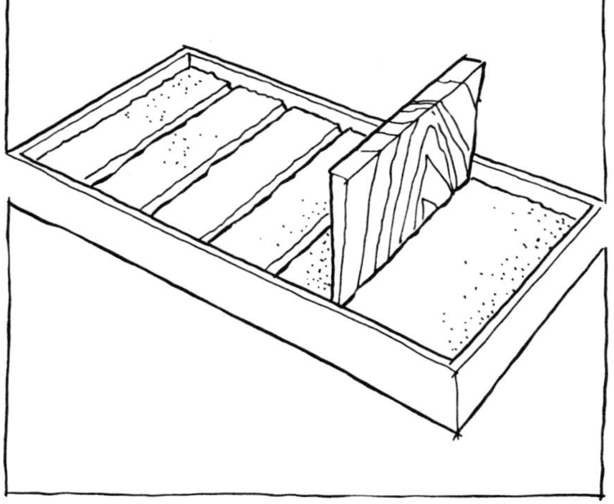

Make seed furrows with board edge.

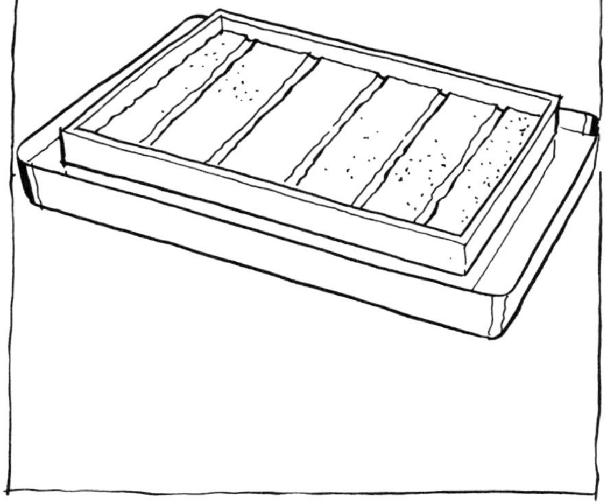

Water the container from the bottom.

ing their uptake of moisture.

Next, to water: water the planted container from the bottom by immersing it in a shallow, flat-bottomed pan half-filled with water. The temperature of the water is immaterial and tap water will do. If excessive chlorine is present in the water, just let the pan of water stand overnight before using.

When the surface of the vermiculite appears moist (this is usually noticed by a slight change of color or wet appearance or can be determined by pressing down with a finger), remove the seed flat or pot from the pan and allow any excess to drain out.

Then cover the flat with a small pane of glass, or if light is not a prerequisite to seed germination, with a sheet of paper or of cardboard.

One useful technique is to slip the flat into a small plastic bag, a freezer bag or even a bread wrapper, and close the open end with a paper clip or tuck it under the flat.

In any case, always remove the covering or take the flat out of the plastic bag as soon as the first seedlings appear.

Once the flats have been seeded, soaked moist, covered and tucked away in a warm spot, they usually will not need watering again until the seedlings have come up and the cover removed. If encased in a plastic bag re-watering definitely is not needed. In the remote case watering is needed, you simply repeat the moistening process by another brief soaking.

However, chances are that if anything interferes with germination it is because of too much water and over-humid conditions. If it appears that the seeded container *is* over-moist and mildew appears, remove the cover and expose the surface to the sun for an hour so that it dries out promptly. As to temperature conditions necessary for good seed germination, in general a moderately warm location (65° to 75°) will suffice. Don't put the flat in the direct sun or on top of the radiator! Where special light conditions or darkness are requisite for good germination, these conditions usually are set forth on the seed packet, and, for many varieties of vegetables, flowers and fruits are noted in the Appendix.

The time required for the seeds to sprout and emerge depends on the kind of plants being grown and the conditions of the seedbed—the moisture, heat and light. Most seeds take from a few days to a very few weeks to germinate. This information sometimes is given in the catalog or on the seed packet.

Undoubtedly you will peek under the cover to see if anything has come up yet. Be patient. But if nothing is happening after a reasonable waiting period, dig up one or two seeds with a toothpick to check on their progress. Only in rare cases will you have to replant.

Once the seedlings have emerged, immediately uncover the container. Don't be tempted to put them in direct sunlight, but in fuller light is all right. These tiny seedlings are quite tender and must be allowed to become acclimated to their new world.

Since you have now uncovered them, you must pay special attention to their need for water. This again can be done by bottom watering or you can keep the vermiculite just barely moist by a gentle sprinkling from a syringe or clothes sprinkler. Don't be tempted to water them under the tap. The lightweight vermiculite, and even the seedlings, can be washed away.

The first leaves that appear are the seed-leaves, sometimes called heart-leaves. This latter term refers to the shape of the cotyledonary leaves which, in those plants that emerge this way, are actually the two halves of the food storage parts of the seed that are thrust up into the light and turn green. Other plants, particularly those with only one cotyledon or entirely different food storage organs, appear above the soil as thin and fragile stalks or spikelike shoots.

In all cases, the first true leaves, more characteristic of the type of plant you've sown, appear within a few days. At this time, your seeds have become true seedlings and are ready for transplanting.

A few comments about seeds and seeding: Sow a few more seeds than the number of plants desired. The weaker ones can be thinned out now. This is done with the least disturbance to the others, by snipping them off at the "soil" line. A pair of thin, sharp scissors is the best tool. Don't pull the unwanted seedlings

Snip off unwanted seedlings.

out, since this will damage their neighbors' root systems.

Save the unused seed, storing it in a cool, dry place for another year. The refrigerator is all right, if the packets are well protected from moisture. Or swap with your neighbor for his unused portion of a different variety.

In a few instances, particularly where seedlings don't do well when transplanted, the seeds are planted directly into the containers in which they later will be set out in the garden. Peat pots are most admirably suited to this, as well as the disk-like peat pellets, which when moistened and swollen to size are sown with the desired seeds. Lacking either of these, paper or styrofoam drinking cups, with drain holes punched in the bottoms, will do. Or you might try an egg carton. The pockets neatly filled with vermiculite will do nicely to hold a dozen seeds.

Peat pots or discs work well.

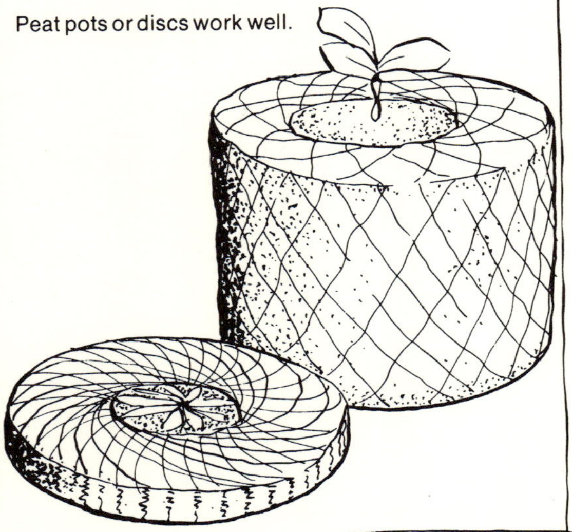

When planting directly into peat pots or other containers, sow two or three seeds per pot. These later will be thinned to the sturdiest one. When these seedlings are large enough to transplant outdoors, simply place the peatpot, seedling and all, directly into the soil of your garden. If you use non-biodegradable containers such as styrofoam or plastic, these will have to be discarded first. Since the objective of seeding directly into the transplant containers is to avoid seedling injury, the peat pots would seem best.

And a final comment: Don't start your seeds too early—six to eight weeks ahead of the anticipated date they will be transplanted outdoors is about right. Old-time Vermont gardeners always started their tomato seeds on Town Meeting Day—which came about eight weeks before the date of the last killing frost—so this allowed just about the right length of time to grow tomato transplants for their gardens.

Transplanting Seedlings

The secret of successful transplanting is to give the seedlings room to grow as soon as possible. If they are left overlong in the container where they germinated, their roots, which develop most rapidly, soon will intertwine. In any case, within only a few days after emergence, the tiny rootlets will have grown so much, that no matter how careful you are, they are bound to be disturbed.

The best time to transplant is as soon as possible after the cotyledons open or within forty-eight hours after emergence. Then, the least damage will be done to the tiny rootlets and the "transplant shock" eliminated.

The seedings are spaced out in larger flats, or are transplanted one to a peat pot or to other kinds of small pots, which should be at least two and a half inches in diameter. Two to three-inch peat pots are fine. Or one can use paper or styrofoam drinking cups. Cut-off milk cartons, the half-pint size, are excellent. But don't forget to put drain holes in your makeshift transplant containers. It is essential that the containers be as clean as possible to avoid disease problems such as damping off.

The "soil" or transplanting media in which the young seedlings will soon be growing now must provide plant food in addition to furnishing support, moisture and warmth. Most transplanting mixtures will contain soil to furnish some of the nutrients or plant food, even though additional plant food may be added later from time to time.

The best transplanting mixture for most seedlings contains equal parts by volume of good loamy topsoil, peatmoss (or leafmold) and sand (perlite or vermiculite). A quart each of loam, peatmoss and vermiculite plus a tablespoon of ground limestone and a tablespoon of superphosphate or two tablespoons of bone meal will be the best.

The secret is to find good rich topsoil loam. Soil from a garden that is doing well should be all right. Loam when rubbed between your thumb and fingers is neither too gritty (sandy) nor too sticky (too much clay). Loam or potting soil often is available from your local florist, greenhouse or garden supply center, as are prepackaged transplanting or potting "soil" mixes.

Commercially-prepared transplant "soils" usually are sterile. If you mix your own, it should be sterilized to be on the safe side. You can do this most easily by baking it in your oven for an hour at 275°F or one and a half hours at 250°F. The soil should be spread out in a shallow pan for thorough heat penetration and well soaked with water prior to baking.

The soil is sterile when the top begins to crust and steam. One way to be sure it has been thoroughly heated is to bury a small potato in the center of the pan of soil. When the potato is baked, the soil is "done."

Some object to sterilizing soil in the oven because of unpleasant odors. If the soil is thoroughly soggy-wet before baking, very little odor is given off, and then only at the very end of the baking.

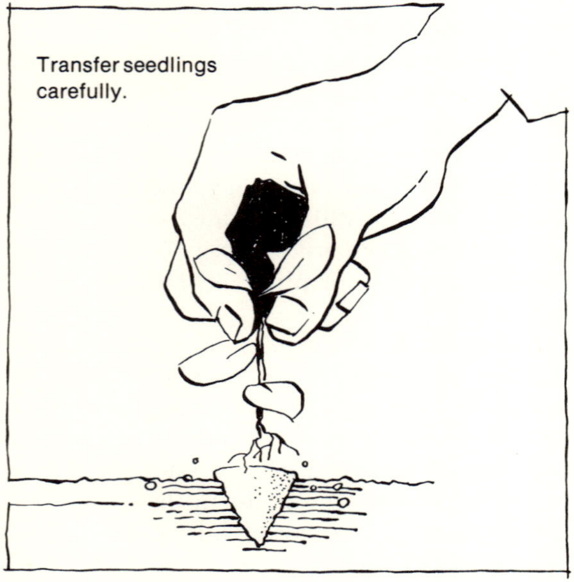

This oven-sterilized transplanting mixture should be well-cooled before use, and if not used immediately it may be stored for a long time safely in tight containers.

The next step is to fill your transplant containers loosely with this sterile medium or transplant "soil." There is no need to press it down this time, since that will be done when firming in the young transplanted seedling.

Tiny seedlings are "pricked" out or lifted out with a small spatula or ice cream stick. Larger ones can be pried gently up and out of the seedbed with a knife or fork. Don't pull them out! Handle it carefully by the leaves, between the thumb and forefinger, so not to break the tender stem. Transfer it to a hole of adequate size to accept the seedling rootlets prepared in the transplant container medium. Depending on the size of the seedling, anything from a pencil to a pointed stick the size of your finger—or your finger—will do the trick.

Lift the seedlings one at a time and transfer them quickly to the prepared container. This will avoid having the rootlets dry out, which would mean the end of the seedling. The depth at which the transplant goes is not important, just so the green leaves are above the surface of the soil medium. Tomatoes are an exception; they can go in as much as an inch deeper. Now, firm the seedling down into the soil gently with your fingertips, pressing downwards but not tightly against the seedling itself. If more potting soil is needed, now's the time you add it. However, don't fill the container to the brim. Level it off a half inch or so below the rim.

Now, give it a thorough soaking, either from below or gently from above, and place the container under the lights.

Plant Care

As the seedlings have turned green, containing their own food-manufacturing chlorophyll, they need more light and a source of plant food nutrients. These you supply by careful attention to their increased light requirements and by regular applications of fertilizer.

The quality of light now is as important as its brightness and the lengths of the daily periods of illumination to provide the best of growing conditions. The windowsill may not provide adequate light for most plants, and it should be supplemented by artificial lighting.

The best arrangement is to use artificial lighting as the sole source of illumination. Cool-white fluorescent bulbs or special plant-growing fluorescent fixtures will suffice. These come in a number of shapes and sizes (see page 38), but a two-tube fluorescent fixture with a reflector and 40-watt cool-white bulbs, plus two

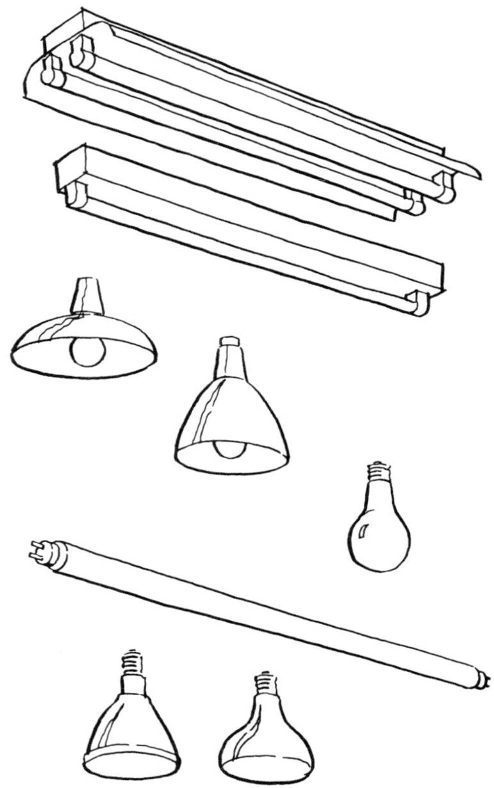

Lights come in many types.

25-watt incandescent bulbs mounted between the tubes has proved to be best for most seedling plants.

The reasoning behind this is that while fluorescent lamps are high in both blue and red rays, they are deficient in the far-red part of the light spectrum. Far-red light energy is necessary for seed germination of light sensitive seeds, for flowering and for general good plant growth. The incandescent bulbs provide the same relative proportions of red and far-red light as does natural sunlight, and also a fair amount of heat. Hence, the happy combination of both cool-white fluorescent fixtures with about one-third of their wattage supplemented with incandescent light.

As the seedlings must be placed fairly close to the lights—about eight to ten inches from the lamps—to receive the optimum amount of light—about 600 to 800 foot candles—some

Height of lights is important, too.

heating is unavoidable. An ordinary thermometer placed at plant level will enable you to watch for overheating conditions. The best temperature for most seedlings is between 60° and 75°F. If it does get too warm, a small fan will keep the plants cooled. But avoid undue air movement, for this as well as overheating will cause the soil to dry out excessively.

The soil medium can be kept moist by frequent waterings. You can prevent excessive drying of the containers by setting them in a pan full of moist sphagnum moss. This also helps keep the air around the seedlings more humid.

How long should you keep the lights on? In most cases, fifteen to eighteen hours each day. Shorter day lengths, as short as ten hours, and cooler growing conditions usually result in sturdier, more compact plants. Longer days, up to nineteen hours of light per day, and warmer conditions will cause more rapid plant growth, but if the light *intensity* is low, the seedlings may become spindly. Tomato seedlings, for example, will grow the best under very bright light, at 65°F, with a long, sixteen hour "day."

The specific combination of light intensity, length of day or lighting period, and temperature varies somewhat with the type of plants being grown. When it comes to flowering, plants are divided into either long-day or short-day classifications, yet there are some that respond well no matter what the day length. However, at this point—of raising seedlings into healthy, vigorous plants ready for our use elsewhere—we are more concerned with vegetative growth and root development. And some plants just grow better under cooler, shorter days, while others such as squash and eggplant like more heat and light.

But all will grow fairly well at 65° to 75°F with the lights on for fifteen to eighteen hours. Plants with special seedling-growth light requirements are given in the Appendix.

Although there may be some plant food in your transplanting media, especially if good, fertile loam was included, it will be necessary to add more nutrients from time to time to fertilize your young seedlings. This is to ensure them adequate supplies of all the nutrients vital to their healthy growth.

You can do this most conveniently by applying a water-soluble fertilizer containing equal parts of nitrogen, phosphate and potash, such as 20-20-20. These numbers refer to the percentages of those three nutrients in the order given above. Drench the containers at every watering with a dilute solution of one teaspoonful per gallon of water.

Commercial water-soluble preparations available include Ra-Pid-Gro, Cangro, Hyponex and Ortho brand, several of which contain trace elements or micronutrients. Some such as fish oil emulsion and kelp preparations, are preferred by organic gardeners. Always follow the directions given on the package, and don't feed them too much or too often. This can be as disastrous as over-watering.

There are no hard and fast rules on fertilizing, just general guidelines and the results of lots of experience. While plants differ in their requirements for plant food elements, for most seedlings a balanced diet of equal amounts of nitrogen, phosphorus and potash is right. The plant food requirements for growing larger plants in containers, flowers, fruits, trees and so forth are given in Chapter VIII in more detail.

Before transplanting the seedlings into their permanent outdoor containers or locations, whether in the garden or decorative containers on the porch or patio, it is wise to acclimate them slowly for a short period of time. A cold frame is most useful for this, but lacking one all you need to do is put the seedlings outside in a somewhat shaded area for an hour or so, each day. Gradually increase their "outdoor" period until they become accustomed to outdoor temperatures and sunlight. If natural shade is not available, contrive some from an old bamboo shade, piece of awning or stiff cardboard. You will need it only for a few days until the seedlings have "hardened" off, so it doesn't matter if it's not pretty.

Herbs always give a touch of spring to the house. While you don't need to grow most kinds as seedlings (they do well when sown directly into the garden soil) a few benefit from such a headstart. And of course, you benefit from having their flavor and fragrance right at hand.

Chives are easily started from seed indoors. Just follow the simple directions on the seed packet. Sometimes you will find chives sold in clumps or bunches in the grocery store. Those with good roots attached can be transplanted into small containers and managed successfully as any other transplanted seedling.

Young chives will be ready soon to go out into the garden, or for you to transplant into their permanent pot for your kitchen windowsill. Seedling chives thrive best under moderately bright light (650 to 1000 foot candles) and when the lights are on for twelve to sixteen hours. The seedlings grow slowly, so be patient, and if you keep them on your windowsill be sure they get extra light.

Another herb that benefits from an early indoor start is parsley. Its seedlings like a slightly cooler "climate" than chives, about 60 to 65 degrees, but needs lots of light to thrive. Parsley seedlings grow slowly, so they can be among the first to sow in your indoor seedbed.

Other herbs can be tried under the lights, particularly the ornamental sage *(Salvia officinalis)*, marjoram, basil and borage, savory and thyme. But these usually are grown as mature plants in containers in your indoor kitchen garden. It's really not necessary to start them as indoor seedlings for your outdoor herb garden, although you might wish to try some to get a headstart on spring.

Crocuses, daffodils and tulips probably are the flowers most used to advance the arrival of spring. Who cannot recall paper cut-outs or crayoned tulips in school? These spring-flowering bulbs and others, including the very colorful and fragrant hyacinths, squills and narcissi as well as grape hyacinths, can be grown easily under lights in your home.

You'll need to plan ahead to bring these spring flowers into bloom inside. Starting in October, plant the bulbs in potting soil directly into their permanent pots or decorative containers. Use moderately shallow containers or 3/4 pots (page 65) and allow the top third of the bulb tips to protrude above the soil. Water them thoroughly and then place these planted containers in a cool (35° to 45°F) dark place, such as in your refrigerator. Infrequent waterings will be sufficient to keep them moist.

From mid-December to mid-January they will be ready to be moved to a slightly warmer location (50° to 55°F), but still kept in the dark. When the flower buds have emerged far enough out of the bulb—you'll have to peek once in a while—place them in the light. Your fluorescent light set-up for seedlings will do, even a couple of 60-watt incandescent bulbs. Give them ten to fourteen hours of light daily. The leaves will turn green almost immediately, and the buds will turn from their pale yellowish color to their brilliant spring hue.

Stretch the blooms over a several week period by not bringing all the containers into the light at once; then, when blooming, keep them cool at night.

After blooming keep the plants in moderate light in a cool place until all the foliage yellows. Now let the soil dry out thoroughly and store the dry bulbs in paper bags until next fall.

CHAPTER THREE

Making Summer Last Longer

The joys and benefits of flowers and fruit need not end with the arrival of cold weather. By providing the proper indoor environment for their continued growth and well-being, many plants, even some fruits and vegetables, can be brought indoors. All that's required are healthy plants that are suited to transplanting into containers, or plants that can be propagated vegetatively.

Many a small seedling that has been raised in the outdoor garden thus can be transplanted successfully into a decorative container and transported indoors, where the right light, temperature and humidity will let it prosper and provide you with beauty, food or flavor for many months to come.

Many annuals, plus some perennials such as marigolds, nasturtiums, heliotrope, chrysanthemums, myrtle *(Myrtus communis),* and of course such tasty herbs as rosemary, marjoram, oregano and the basils, come to mind as plants that can easily be started from seed in the garden and then transplanted into pots. And don't forget tomatoes!

The best way to do this is to transplant the seedlings into pots early in September, then bury the pots up to their rims, right back into the garden soil. Water well and provide some temporary shade if the spot is overly sunny. Then in early fall, lift the pots, and clean off adhering soil with a damp rag, and bring them indoors to a well-lighted place. They can stay right in the same pots, or the pot slipped inside a slightly larger, more decorative container, or they can now be repotted into the new containers.

Mature plants of an appropriate size also can be potted outdoors ahead of time, in preparation for their move indoors before winter sets in. Typical of these would be young plants you started from cuttings and set out in the garden earlier, or larger plants previously grown from seed. Some examples of the former are Shrimp Plant, chrysanthemum, gardenia, myrtle, geranium and all the ivies. Examples of the latter would be Lemon Verbena, rosemary ageratum, Flamingo Flower, the bromeliads, Norfolk Island pine, mint, parsley and chives.

For either seedlings or mature plants, use a good potting soil mix. Especially for large containers consider a lightweight soil mix. Commercial potting soils are excellent, and if made with vermiculite or perlite and sphagnum moss will weigh much less than the conventional sand plus soil plus leaf mold potting mix. Chapter VIII gives hints and "recipes" for potting soil mixes.

When larger plants are to go indoors, some advance preparation may be useful. Make cuts with a sharp-edged trowel or small spade down into the soil around the plant. The cuts are made in a circle slightly smaller than the pot into which the plant will go. For some varieties, known to have a deep tap root, undercutting also may help. The plant then is watered, thoroughly soaking the soil, and left undisturbed right there in the garden for a week or two before being lifted out and potted. This allows the root system to recover from its "pruning."

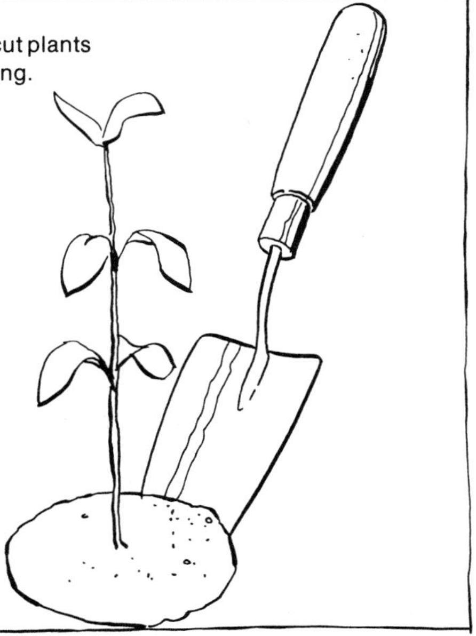

Circle-cut plants for potting.

During this time, if it's needed, you can give the plant a light pruning, removing unwanted branches, excessive foliage, or just a general shaping up.

After potting, and just prior to bringing them into the house, inspect your plants for insects. You don't want aphids, mites or white flies hitchhiking into your home. If insects are noted, you can get rid of them with one of the safe garden plant insecticides sold in garden supply centers. Be sure to read the label and use only the materials recommended for the kinds of insect pests you have and for the kind of plants they're infesting. If you're organically minded (or not) a very dilute solution of yellow naphtha soap works well. Dissolve a tablespoon of grated soap in a gallon of warm water and when this is tepid or cool, syringe it on the plants. A garden hose spray attachment is quite useful for this. Lay the plants on their sides before any spraying so you can more easily get at the bottom sides of the leaves. Once your seedlings or mature plants are safely in their containers, you can bring them inside.

The environment of your house usually is more suitable to the comfort of people than to the welfare of plants. So now you must consider what proper plant growing conditions you'll need. The basic, two-tube fluorescent fixture that you used last spring to start seedlings indoors is an excellent source of the extra illumination outdoor plants need when growing indoors.

Get rid of insects before bringing plants indoors.

There are many other sorts of lights and lighting arrangements, and these are discussed in detail in Chapter IV. While most plants will do well with moderate light intensities and a fourteen to sixteen-hour day with warm (70° to 75° F) daylight periods and perhaps a bit cooler at night, some plants do have specific requirements for their best growth. These, too, are noted in Chapter II and in the Appendix.

Infrequent soakings are much better than over-frequent sprinklings. A good way to do this is to immerse the pots in a tub of water—the kitchen sink or the bathtub would be fine—until all the air bubbles have stopped coming up from the soil. Then you let them drain—drip dry in effect—before replacing them in their lighted spot.

Where humid air is a necessity, the pots can be placed in a moist sphagnum moss bed inside a larger container. Teflon-lined baking pans are excellent for this, although most garden suppliers have rectangular plastic trays about two inches deep. Pans of water are better than infrequent spraying of the foliage to maintain humid conditions.

Newly-potted plants coming into your house

from outside will respond well to a modest helping of balanced plant food. Use according to directions if you're feeding a commercial preparation; otherwise, water once with a solution of one tablespoon of 20-20-20 per gallon of water. Feed again in three or four months. If your potting soil mix contains rich loam or fertilizer (check the label on the package) omit the first feeding.

Vegetative Propagation

Once you may have thought that starting plants from slips and cuttings was an art belonging exclusively to private gardeners of Scotch ancestry—but it's not so today. You can propagate your own plants easily from rooted cuttings with the aid of simple supplementary lighting and some attention to proper air humidity and temperature.

Vegetative propagation refers to reproducing a plant by the use of slips or cuttings, divisions, ground layering, air layering, offsets and grafts. Most of the ornamental shrubs, and many of the more succulent perennials and herbs too large to be potted before their entrance into your house can be duplicated by rooting small pieces of leafy stem. Some are propagated by forcing roots to grow from single leaves.

Although incandescent lights often will suffice, your simple low-wattage fluorescent light set-up is preferred. A two-tube, 40-watt unit will provide enough light for several containers of rooting slips or cuttings. Most cuttings of succulents, sun-loving herbs, cacti, roses, geraniums, and many other softwood and hardwood cuttings prosper when kept about six inches from the lights. More shade-tolerant plants such as begonia, African violet, ivy, philodendron, the gesneriads and conifers, do better when ten to twelve inches from the lights. Most evergreens are sun-loving and will do as well under more intense light, but they must be kept fairly cool. All cuttings will root well with fourteen-hour days, but do best with the lights on for sixteen hours each day. Day length is more important than light intensity.

Two things will help greatly in maintaining the cuttings in a very moist atmosphere: a closed but transparent container and a rooting medium that holds water well.

All sorts of glass and plastic containers satisfy the first prerequisite—glass-covered pots or jars, plastic refrigerator dishes or any of those clear hard plastic boxes designed for loaves of bread, refrigerator food storage (especially those for bricks of butter or cheese), even plastic sweater boxes! Those with solid bottoms should have drainage holes provided. This you can do by drilling a few small holes or carefully punching through with a heated icepick.

The following are recommended and used, alone and in various combinations, for rooting media: sand, perlite, milled or screened sphagnum moss, peatmoss, shredded sugar cane and vermiculite. Some softwood cuttings or slips are easily rooted just in a jar of water. For acid-loving plants such as azaleas, a mixture of sand and acid peatmoss does fine, but for most all other cuttings I've found vermiculite to work best. Vermiculite doesn't need to be sterilized, and its holds much more water than any of the other media. About its only drawback is that it can become over-soggy; but only if the drain-holes are plugged.

Fill the container to within a half inch of the top with vermiculite. A number of small 3/4 clay pots, filled with vermiculite and immersed in a bed of moist vermiculite or sphagnum moss are useful if you're handling a number of cuttings from different plants at the same time.

Wet the vermiculite carefully but thoroughly, so that water drips through the drain holes. Let the excess drain out, and it's ready for your cuttings.

The best temperatures for getting roots to develop on cuttings are between 70° and 75°F. If your room is cool, some supplementary heating may be required to keep the medium warm. The best is a thermostatically-controlled soil heating cable to provide gentle bottom heat. Others have rigged up ingenious setups including water baths on the radiator, light bulbs under the flats, even an electric heating pad—although here one would have to be very cautious to keep it dry to avoid electric shock hazard. Electric soil heating cables are available from greenhouse and garden supply centers and can be used again in the spring in outdoor hot frames.

Cuttings usually are divided into two categories, softwood and hardwood. Although the terms denote the kind of stem being cut, most softwood cuttings are taken from the active vegetative growth in spring or summer, while hardwood cuttings are those taken during the dormant period—usually in the fall after the leaves have dropped but before cold, freezing weather sets in. Other minor categories of cuttings include root cuttings and leaf cuttings, which will be discussed later. Acacia, aralia, azalea, blueberry, daphne, flowering maple, gardenia, geranium, hydrangea, ivy, Norfolk Island pine, peach, philodendron, quince, and rhododendron are good examples of plants easily propagated by cuttings.

Softwood cuttings are the easiest to root, and can be taken any time during the spring or early summer while growth is active. Best cuttings come from the younger, side branches.

Most cuttings are three to four inches long, but with lights and controlled humidity, shorter cuttings are easier to handle.

Use a sharp knife. Make the cut just below a node or leaf bud. Strip off the lower leaves, dip the cut end in hormone rooting powder, then insert this end in the moist vermiculite. Prepare a small hole first with a pencil or pointed stick. Following the hormone powder treatment, the cutting should be placed at once into the vermiculite. Insert the cutting carefully so not to rub off all the rooting powder; then firm the vermiculite around it. As soon as all the cuttings have been inserted, cover the container (see above) and place it under the lights. In about two to five weeks, most softwood cuttings will have rooted. It's hard to keep from digging them up to see, but be patient.

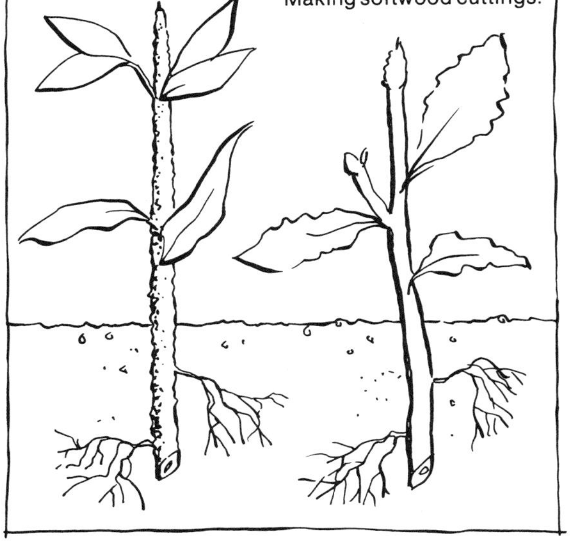
Making softwood cuttings.

If for any reason there is a delay between taking the cuttings and inserting them in the vermiculite, keep them covered up in moist sphagnum moss or burlap. Never leave them out in the sun nor place in standing water.

Two useful tricks in making cuttings—one old and one new—are to use an inverted glass jar over a few cuttings in a pot or pan of vermiculite; the other, to root the cuttings inside a small closed plastic bag half filled with moist vermiculite.

Hardwood cuttings, made best just after leaf-fall, usually take much longer—up to a year sometimes—before good roots develop. However, with proper lighting most will root much sooner than this.

Cut branch tips of moderate size—up to the diameter of a lead pencil. Be sure to include at least two leaf nodes, so the actual length of the cutting will depend on how far apart the nodes are. Make a slanting cut at the top. Discard the more succulent tip portions. Dip the bottom end in hormone rooting powder and insert it, as with softwood cuttings above, into the vermiculite. Cover and place under the lights.

As some hardwood cuttings are necessarily long so as to include two buds, a rather deep propagating container must be used, or use the pot with the inverted glass jar technique.

Don't disturb the cuttings by continually lifting them to see if roots are developing. Wait until the buds swell and the leaves grow. If there are new roots, the leaves that open will continue to stay green; without new root development, the leaves may yet open but soon will wither and die. As soon as good rooting has occurred with both softwood and hardwood cuttings, the cuttings may be transplanted. This can be directly back into the garden or into suitable containers, with the new young plants

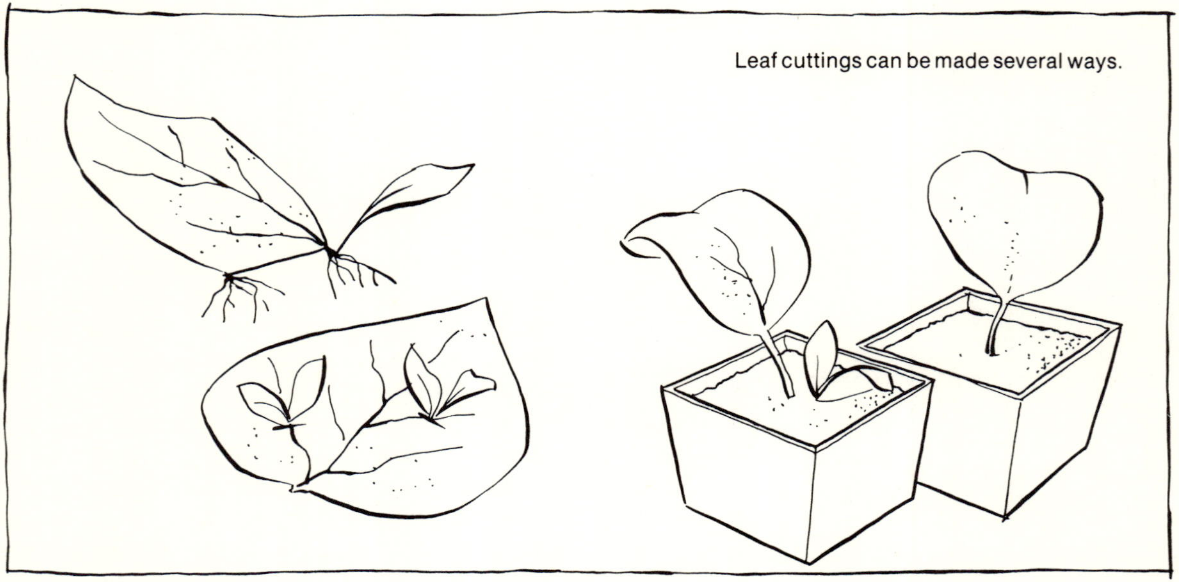

Leaf cuttings can be made several ways.

allowed to continue to grow under the lights.

Some advanced planning is necessary if you wish to have new plants to bring into the home, and you're propagating them by rooting cuttings. Hardwood cuttings, taken necessarily in the late fall and slow to root, often need many months to develop into desirable container plant specimens. Softwood cuttings, taken from actively growing portions of the stems, will develop sooner. Many of them will grow into satisfactory container plants for your use and benefit by the time fall comes. With many species of plants, rooting cuttings is a very satisfactory way to extend the benefits of summer.

Leaf Cuttings

Another method of propagating plants vegetatively is to root new plants just from cut-off leaves. Many fleshy-leaved plants will send out roots from an entire, isolated leaf or even portions of leaves. This is because the leaves themselves contain a good store of food. Many examples of plants that can be rooted easily this way are listed in the Appendix and include African violets, gloxinias, Rex begonias, sedums and the other succulents.

Several techniques work well: inserting the petiole (leaf stalk) in the moist vermiculite of your propagating chamber, or using leaf sections which are "planted" in the vermiculite, or by first cutting through the main veins with a sharp knife or razor blade, then laying the whole leaf, top-side up, on the surface of the vermiculite, pegging it down tight to the moist rooting medium with toothpicks.

Once roots start, new leaflets will arise, and you can then separate these new plants, potting them individually and returning them to the lights. The light and humidity conditions are the same as for softwood cuttings and noted earlier.

Root Cuttings

Yet another technique for propagating new plants from old, is to produce new rootlets and top sprouts from segments of the roots cut from the old plant. Blackberries, raspberries, and the other cane fruits can be multiplied easily in this manner. Others are listed in the Appendix.

Pick the most vigorous plants to give up some of their roots. Select roots at least as thick as a soda straw but not much over the diameter of a pencil. Cut them into two- or three-inch long segments. Bury these one-half inch deep in the moist vermiculite of your propagating box. Light will not be necessary until the shoots pop up.

Divisions

Some plants, particularly perennials such as columbine, strawberry, clump-type bamboos and the bromeliads can be divided into smaller plants, each of which already will have roots and can be potted in a suitable container.

Select the most vigorous growth, usually around the perimeter. Of course, you must first dig up the old plant early in the fall as soon as it becomes dormant. Then quickly and in the shade, pull the clump apart into desirable-sized segments. You will find that a spading fork or hand fork is an excellent aid in separating the tangled roots and crown sections. For tougher, more woody perennials you may have to resort to the use of a knife or hatchet to sever the sections. With the exception of most evergreens, last season's top growth usually is cut back to within four to six inches of the ground line.

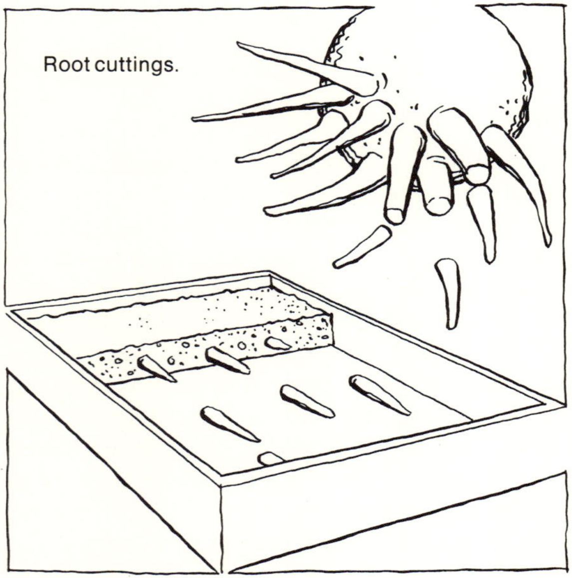

The best segments are then replanted, and this can be right back into the flower bed, but if you wish to force fruit or bloom with the plants indoors, pop them right into their containers.

Whether or not these particular new plants need a resting period (see Appendix) before breaking dormancy and growing again, will determine whether you now store them for a quiet rest in some out-of-the-way place, or bring them under the lights so to enjoy their benefits once again.

Some perennials have single tap roots and of course are better propagated by other means, but some of the perennial plants that can be divided in this manner are noted in the Appendix.

Bulbs, Corms, Tubers and Rhizomes

Daffodils and the others of the narcissus family—some lilies and onions—and of course chives—produce small bulblets next to the mother bulb. To start new plants of these sorts, dig them up in late summer or early fall, and pot up those bulblets that most easily can be broken away. Gladiolus and other plants that grow from corms can be lifted from their pots or from the garden after blooming and after the leaves have yellowed and started to wither. You frequently will find that a large new corm has grown on top the old, perhaps one or more smaller corms and frequently several tiny cormels. All these can be coaxed into new vigorously flowering plants given the proper growing medium, rest periods, then light and warmth. You even can divide a large single corm by cutting it in halves, thirds or even quartering it. If you do this, dust the cut surfaces with powdered charcoal or sulfur to help callus formation.

Corm with bulblets.

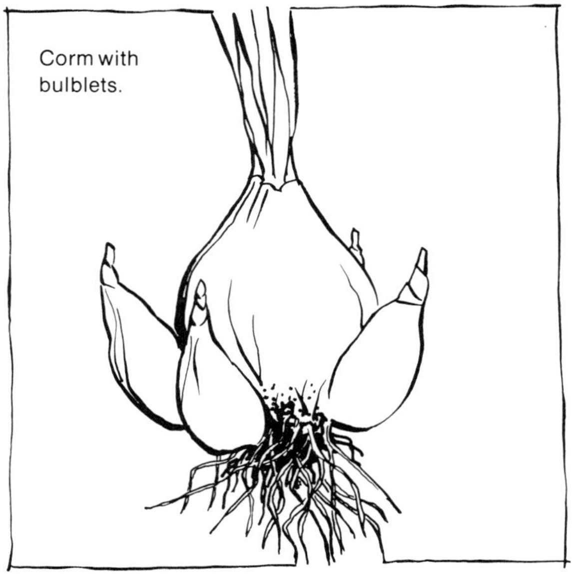

Dahlias, tuberous begonias and sweet potatoes reproduce by tubers. These actually are enlarged roots chock full of stored food. You usually would dig dahlia tubers in the fall after the first frost has browned their leaves. Let them dry. Then with a sharp knife cut off the new tubers, being very careful that each has a piece of the mother plant's crown and at least one dormant bud or "eye" attached. These then should be stored in a dry place against next spring's planting, but I can see no reason why later on you might not like one of the new dwarf varieties inside your house in a container under lights.

Tuberous begonias usually are divided in early spring when the new buds start to emerge. You cut the tuber into sections from top to bottom, so that each piece has at least one growing bud. Dust the cut surfaces with charcoal or sulfur and start them under lights.

Iris, running bamboo and white potatoes come to mind as plants with rhizomes, although the latter are more commonly referred to as tubers. You divide iris with a sharp knife after lifting the plants from the soil in late summer. Save only the newest, most vigorously-growing outer rhizomes, trim the leaves to three or four inches above the ground line, then replant or pot. Just barely cover the top of the new rhizome section with potting soil.

Dividing tuberous begonia.

Runners, Stolons and Offsets

Still another method that nature has for vegetative reproduction is for the mother plant to grow new baby plants at its side—or perhaps at some distance, at the end of an umbilical-like runner or stolon. Some good examples of these are the strawberry begonia (in fact, almost any of the true strawberries), the spiderplant, walking iris and piggybacks. As in the case of lilacs, raspberries and Kafir lily, these offset plants are called suckers. Some, such as amaryllis, develop from offset bulblets.

In any case, these are noted in the Appendix, and all can be divided easily and propagated vegetatively to your benefit. The easiest way to go about this is to isolate the young plantlets, then plant them, either directly in the propagating case or in their new container homes, of course providing the proper light and moisture.

Layering

One special way to reproduce plants is by grafting a bud or stem of a desired plant upon another. But since this is not useful to make new plants to bring into the home, but rather to change the characteristics of already existing plants, it is a subject that is discussed elsewhere.

However, there is one additional technique of producing additional, new plants from an older plant: the process of growing roots on a section of stem or branch, with the branch still attached; then cutting it off and voilà, a new plant! And you still have the old plant, too.

There are two methods of doing this layering; one being the notching of a branch and then bringing and holding that portion of the branch in close contact with the soil while it takes root. You can propagate forsythia and rhododendron quite easily this way.

This often is a slow process with some species, but you can have a new plant to bring into your indoor container garden that otherwise might not be available. All you do is pick out a low, vigorously-growing branch, one that easi-

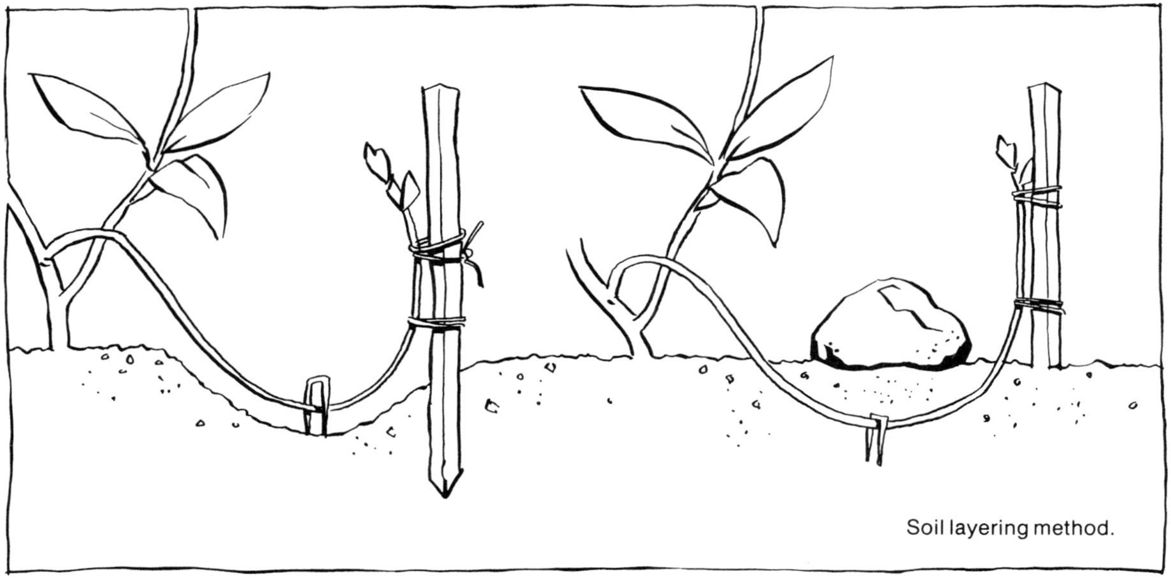

Soil layering method.

ly can be bent down to the ground. About a foot or so from the tip, make a small notch on the underside with a sharp knife. Use a slanting cut and cut not quite half-way through. Wedge a pebble or toothpick in this cut to hold it open and then bury this portion of the branch two to four inches deep. Cover the cut portion of the branch with good potting soil mix and staple the branch down with a giant wire staple made from a wire coathanger—or better weight it

Air layering technique.

down with a small rock. If desired, you can tie up the tip end on a small stake so that the leaves are well up in the air and sunshine. When the roots are well developed, sever this new plant from its parent and pot it up in an appropriate container.

The other method is called air layering. This you do by making a slanting cut not quite half-way through the stem of a shoot or branch. Make the cut just below a bud or node, wedge in a small pebble or toothpick to keep it open, dust the cut lightly with hormone rooting powder. Next, wrap a generous handful of damp sphagnum moss around the stem and bind it up securely in clear polyethylene sheet plastic. Tie with string both above and below the sphagnum-wrapped cut.

When roots are visible in the sphagnum beneath the plastic, the new plant can be severed from the mother plant, and potted in your favorite container. You can shorten up old bare-legged rubber tree plants by this method!

While some of these methods of getting plants together to bring into the home may seem complicated, they're really not. It's fun to try, and chances are, by providing the best conditions for these summer plants to develop and grow, you will be blessed with beautiful flower and foliage, fruit and tasty herbs long after all other outdoor plants have gone by.

CHAPTER FOUR

Summer Indoors Year-Round

I've mentioned grandmother's ferns and geraniums before. Keeping potted plants in the house is nothing new, but yesterday's house plants now have become today's indoor container gardens. The use of artificial illumination has moved the plants off the windowsill and into any room in the house—including closets, cellars and attics!

Of course you still may grow plants on the windowsill or near it, but when you supplement the natural light with additional illumination your ability to grow plants will more than double.

Now it is possible to grow indoors in your home or office, plants that prospered only in a florist's greenhouse or in a sunny garden—things such as your own salad garden, tomatoes, even orchids. Previously difficult-to-grow African violets and gloxinias, now under lights, grace many a home and business establishment. You can have a year-round supply of herbs and salad greens growing in the kitchen, fruit trees in the foyer, summer flowers in the living room.

Once you get started on indoor, year-round gardening you will discover the plants or flowers most to your liking. You might even become a hobby gardener, specializing in certain kinds of plants: a collection of African violets, scarce gesneriads or succulents, perhaps miniature roses or even orchids, or miniature bonsai or bottle gardens. Or perhaps you'd just like to concentrate on salad greens and herbs. In any case, your efforts will be rewarding and success easily obtainable with a good indoor-gardening light arrangement and careful attention to humidity and plant food.

Windowsill Gardening

Perhaps the easiest place to start your indoor garden is on or near a sunny windowsill. Its advantages are increased with the arrangement of suitable lights to supplement both light intensity and the hours of daylight available from nature.

A window on the south side of the house is preferable, and it will be the happiest of circumstances for the plants if it is a bay window. With a bay or bow window you'll get additional hours of early morning and late afternoon sun.

Next best exposure is facing toward the east. West windows get less light and north-facing ones perhaps no direct sunshine at all. You will be able to grow some plants in any of these windows, depending on their light requirements (see Appendix), but to achieve the full benefits from indoor gardening you'll be better off with some supplemental lighting. This is especially true for windows receiving little sunshine.

When you grow plants in sunny windows you'll notice that their flowers and foliage soon will be facing the sun—"looking" out the window. Since you want to enjoy them from inside the room, you should turn the containers every other day or so and the plants will grow more uniformly.

Some hazards of windowsill gardening easily can be avoided if anticipated. These are, drafts, sudden and extreme drops in temperature, and excessive drying conditions.

Most plants don't need fresh air, but a change of air is welcome. However, cold drafts coming through the window can be detrimental to sensitive and heat-loving plants. Drafts also accelerate the soil's drying out. It is important to prevent them, particularly during cold winter spells, and this you can do simply by caulking or taping up the cracks. A more permanent solution would be to weatherstrip the window.

Although most plants when growing outdoors are used to a moderate drop in air temperature following sunset, the temperatures next to a window often plummet excessively during very cold winter nights. Usually this can be anticipated and suitable precautions taken.

You can move the plants away from the window or draw the drapes. This may be quite a task if you have many plants, and next to impossible with containers of plants on brackets or shelves across the glass. The easiest way to provide protection for your plants is to cover them with cloths or small tents made of newspaper. These you would remove next morning when the room again warms. One trick grandmother used was to tack sheets of newspaper across the glass. Today, a piece of clear plastic of suitable dimensions would be taped or stapled there.

One caution about "fresh air": If your indoor garden is in a kitchen with a gas stove or other gas appliance, be sure there are no faulty fixtures or pilot flames that blow out, allowing gas to escape. There may not be enough fumes for you to detect, but many plants are supersensitive to gases and could die quickly.

Humidity and soil moisture can be maintained properly too, with some advance preparation. Overheating from too much sun or too hot a room or being too close to the radiator, generally is not a problem. What *is* a problem is the rapid and excessive drying of the air, and the soil as a result. Supplemental, artificial lighting may generate enough heat so that plants that are too close will dry out.

Unlike plants in a natural setting with their roots in the soil and only their tops exposed to the sun's heat, plants in pots often are more vulnerable to drying. They lose moisture through the walls of the pot itself. I mentioned setting the pots in moist sphagnum moss earlier (page 19) as a precaution. Another way would be to set the containers in pans of water. The new teflon-coated baking pans are excellent for this, not only colorful but rust-proof. Rectangular plastic trays are good, too. A pan of water on the radiator also helps keep up the air humidity. Automatic watering devices are discussed later in Chapter X. These, too, are very helpful to keep your windowsill containers moist.

Some indoor gardeners feel that syringing or spraying the plants' leaves helps raise humidity, but I doubt that this is very helpful or effective as far as raising the humidity goes.

Properly arranged and with ample consideration to all of the plant's environmental needs, a windowsill garden can be a most attractive spot in your home. And it may be a good place to begin at indoor gardening, for only a minimum of light equipment is required to supplement available sunshine.

Although the quality and quantity of extra light needed varies with the kind and species of plants you're growing, and obviously with the amount of natural light available, a few simple rules apply. Most seedlings require higher light intensities and longer days. Foliage plants (and most herbs and vegetables fall into this category) do well only with supplemental fluorescent light. Flowering plants need the more balanced light from incandescent bulbs or combinations of incandescent-fluorescent supplemental lighting. Even better suited for this are the new plant-growth lamps.

Most of your plants will prosper from an additional four to six or so hours of supplemental light so that they have fourteen to sixteen hours of total "daylight." For flowering plants this will vary as to whether they are short- or long-day types.

As far as the intensity of the supplemental lighting is concerned, this you can judge most easily by the appearance of the plants. If wan and spindly, they need brighter lighting. Very few plants can get too much light, except some of the shade-loving types. Crinkled or bunched very dark green leaves may be symptomatic of too much light (or perhaps mite injury). White "burned" areas also can be caused by having the leaves too close to the lamps. Your solution of course is to move the plants a little farther from the lights.

Indoor Light Gardening

The basic light setup consists of a standard 48-inch, two 40-watt fluorescent tube reflector fixture equipped either with two standard daylight (warm or cool) white tubes plus extra incandescent bulbs (see page 17), or it may have one standard tube plus one plant-growth bulb (such as Gro-Lux, Plant-Life, Plant-Gro, Natur-escent, or Vita-Life) or two of the latter. Some come as eighteen-inch, 15-watt tubes, and twenty-four-inch 20-watt tubes as well.

There are of course many other sizes and shapes of lighting fixtures such as rectangular flat fluorescents, circular or ring fluorescents, reflector (R) or parabolic aluminized reflector (PAR) incandescent bulbs useful for spotlighting plants, and high-intensity very high output (VHO), and super high output (SHO) fluorescents. One example of the latter is named Power-Groove.

All these have their special uses, mostly to serve in decorative plant-growing lighting fixtures or to increase light intensities where especially needed.

The standard setup is the same as the one you use to start seedlings in the house. Since we're concerned with providing regular periods of "daylight" to the plants in the house, an automatic timer is an invaluable addition. With one you'll no longer need to rely on memory to turn the lights on or off when it's time to do so, nor to rely on chancy sunlight.

For a simple indoor light garden, suspend the light fixture on chains or on pulley ropes, so that it will be about twelve inches above most of the leaves of the plants you're growing. Set the timer for fourteen to sixteen hours a day, depending of course on the plant's individual requirements. And now your indoor container garden is all set to enjoy "Summer" year 'round.

Indoor Garden Plants

If you're new at indoor container gardening, or perhaps never have gardened at all, or even if you're already well into it, you're probably deciding what plants to grow or what new and challenging plants will give you pleasure.

While almost any plant can be grown indoors, at least some of the time, if you provide it's proper environment, many can be grown indoors year 'round. The kinds of plants that come to mind are flowering plants and bulbs, foliage plants, cacti and succulents, flowering and fruiting shrubs and trees, vegetables and herbs.

Indoor plants are grown singly or in groups or arrangements. You should consider both your desires and the needs of the plants. Are they decorative or functional, light-loving or shade-tolerant? Do you want a single plant as a focal point or a collection of culinary herbs?

In the case of the former, you would give special emphasis to the appearance of the container and perhaps use special floodlighting or a handsome lighting fixture. With a kitchen herb garden the containers would be more functional and the benefits derived from the plants both decorative and useful.

If you plan to grow a number of plants together under the same light setup, you would select plants of similar growth requirements, those that would be adaptable to and do well under the same "daylight" period. By placing plants at various distances from the lights you can grow plants with differing light intensity requirements together in the same area. For example, orchids and African violets will do well together at the same light level, as would a kitchen garden of salad greens and leafy herbs. On the other hand, if your livingroom flower garden included gloxinias and ivy, the gloxinias would be centered under the lights and the ivies allowed to cascade off to the sides.

To plan your indoor garden for easy care and best results, select plants that not only fit together well as to appearance or color but those that will do well within the bounds of the plant growth environment you have provided. Many books and references provide detailed information as to what plants grow well indoors, but information on their proper indoor plant growth environment is hard to find. So we have assembled in the Appendix a selection of the best plants and their growth requirements.

Indoor Salad Gardens

Nothing is more delightful perhaps than a tasty salad made from your very own home grown lettuce, cucumbers and tomatoes. But in the middle of winter? Most certainly. Then it's particularly appetizing because you grew it yourself, there in the kitchen under a simple light setup.

Many delicious vegetables can be grown indoors under lights, but perhaps the easiest and most rewarding are the salad vegetables. There are many places to do this: in a heated garage, the cellar or attic or the kitchen itself, if there is room available for the light setups. Even a taste of lettuce can be grown on the windowsill.

Endive, leaf lettuce, spinach, radishes, onions, green peppers, cucumbers—even tomatoes—will do well under your basic light setup. For peppers, cucumbers, and tomatoes I would suggest one of the new, midget hybrid varieties. These three plants require lots of warmth and light, so they have to be grown separately from the leafy "salad greens" which do best under moderate light and cooler conditions.

While you probably would grow pepper or tomato plants one to a container, and maybe two or three cucumber vines in the same pot, many of the other vegetables could be planted and grown right in the same pot, box or tray. A windowsill planter box makes an ideal container for a miniature salad garden. Several different kinds of leafy vegetables, even midget carrots, can be started from seed and grown to maturity in small rows, all together in the same container under your basic light setup. More produce could be grown in larger flats or boxes under a number of light units elsewhere in the house.

You can grow chives or parsley along with the salad vegetables. These do better in their own containers, but can be placed under the same lights as the box of lettuce and radishes. A salad garden is as easy to grow as raising marigolds or geraniums, and besides looking as nice it tastes better!

Indoor Herb Gardens

Aside from flowers and foliage plants, perhaps more people have grown a pot of chives inside than any other kind of plant. They do well on almost any windowsill providing moderate light. However, with a basic light setup you can grow indoors many other tasty and fragrant herbs.

Where you grow them simply depends on the space you can make available. A kitchen windowsill with supplemental lighting hidden under the valance, a discarded tea-tray wheeled into a corner, even a kitchen cupboard, without doors but with an overhead fluorescent fixture installed, all make excellent choices. You use what you have and what you can adapt.

As far as the herbs themselves go, it doesn't matter if they're annuals or perennials, but you are better off to stay away from plants such as dill that grow rampant and are over-large for kitchen quarters. Start with parsley, chives, savory and borage. Or buy a grow-it-yourself kit containing a collection of easy-to-grow herb seeds, the pots, potting soil and all.

Although chives can be started from seed,

they are more satisfactory when grown from clumps of plants available from greenhouses or even from the grocer's vegetable bin. Started from purchased plants, or those brought in from the garden, parsley thrives under well-lighted but cool conditions, though you can start it from seed as well.

Other indoor herbs include sweet basil, borage, chervil, marjoram, sage and bay. Bays almost are shrubs and like shade. They can be grown in a separate container off to one side of the lights.

You might like to try watercress or mint. Both like bright light, although mint will tolerate low-light conditions. But both require a great deal of water. Ways to provide this are to use an extra amount of perlite or vermiculite in their potting mix, or to keep the pots in saucers or trays filled with water. Otherwise daily soakings definitely are required. Both plants respond well to automatic watering systems (see page 91). Many other kinds and varieties of herbs that grow well in the house, especially under artificial illumination, are noted on page 56 and in the Appendix.

Your herbs will last for quite a while, since you'll be snipping off only a leaf or two at a time. Or you can harvest whole plants if need be. Many of the herbs can be stored in the freezer or by drying, so their benefits extend over even longer periods of time.

Once started on year-round indoor gardening, you will want to add more light setups and expand your repertoire of plants, perhaps to include some that later can be moved to the outdoors, to the porch, patio or garden.

CHAPTER FIVE

Switching Seasons

A great deal has been written lately about container gardening for the patio or outdoor courtyard. One of the great virtues, of course, is the mobility of the plants, which can be moved here and there as you wish to accent this corner or that. Often neglected is the fact that you also can move containerized plants from inside the house to the outside balcony, porch, or patio when the plants may need a bit of sunshine or when you'll benefit most from their outdoor appearance. And of course, you can move them right back inside if need be.

You can practice a sort of indoor-outdoor gardening, switching seasons as you desire, keeping beauty where you want it—indoors when conditions outdoors are adverse for you or the plants, outdoors when that situation is where the plants and you like to be. You can take advantage of the best plant environment. Indoor plants can go outdoors for rest and recuperation, or you can bring outside your about-to-flower annuals and bulbs, or perhaps the flowering shrubs that you gave a headstart indoors under the lights.

Plants That Can Be Moved

Many plants prosper because of this transportability. Many perennials, especially flowering or fruiting shrubs and miniature trees, need the optimum light available out of doors to grow best; others require a resting period to recuperate from their light-forced indoor flowering. Moving them outdoors is one way to accomplish this.

An example of how dwarf trees often need this switching of seasons is that citrus really prosper, after a winter indoors, by going outside in really warm weather. Apples, as another example, are kept in their containers out of doors most of the time and then can be moved indoors when about to blossom. For fruit production, they then must be hand-pollinated.

Most deciduous fruits now are available as dwarf varieties, and are very adaptable to container gardening. Unlike the citrus fruit, however, they rarely are indoors for more than a brief period of spectacular bloom. Low humidity and low light intensities cause a sort of artificial autumn to occur, and their leaves will start to fall prematurely. Of course, you can raise the humidity and increase the lighting, but even the smallest of the deciduous dwarfs is too large to get really close enough to the lights. High wattage reflector bulbs are some help in offsetting this premature leaffall, and also are very useful to spotlight their flowering beauty.

But even if adequate light is available for their good indoor growth requirements, all deciduous fruits (apples, apricots, cherries, peaches, pears, plums) need a chilly dormant period. So, after they bloom indoors, move them outdoors for a while until leaffall, then into a protected out-of-the-way spot where they can go through their cold winter sleep.

Those not "winter hardy" will need protection. In cool climates they can be mulched and wrapped with a pile of straw over the container to keep the roots snug, and burlap around the limbs to ward frost from the buds. In colder areas the more sensitive apricots, cherries, peaches, pears, quinces and plums may be moved into a protected garage or shed.

Most of the other perennial woody shrubs and trees stay right out on the patio during the cold winter, unless of course you wheeled them inside to continue enjoying their benefits indoors. Those left outdoors in cold, snowy, climates often will benefit from having their containers mulched against the cold, and (with some species) their branches wrapped snugly. This is particularly true for roses and some of the more delicate evergreens. The latter usually are wrapped up in winter to prevent leaf injury from chilly, dry winds.

As for the plants that prosper from an outdoor vacation, miniature roses and orchids come to mind. Roses rest during late fall. After a winter indoors you can bring them outside for the summer, placed on the patio in their containers or sunk, container and all, in plant beds or the garden. Come fall they take their resting period in the garden, or if repotted, in a cold frame before being brought back into the house. This gives them a good, safe rest, and they're readily accessible to bring back into the house before Christmas.

Orchids, too, particularly the cypripediums, will benefit from being out-of-doors during the summer. African violets, gloxinias, hanging baskets of trailing sedums and begonias will thrive in somewhat shady outdoor spots.

We often think of plants that can be brought indoors and out as the long-lived perennials in portable containers, forgetting that many of the annuals, bulbs, and herbs and vegetables can be moved about to suit our fancy. Daffodils, for example, after their in-the-dark, pre-bloom conditioning, can be located about the patio. If a late frost threatens, inside again they come. Annual flowers and some perennials, started into bloom under the lights inside, can be moved outside as weather allows.

Herbs growing in a box planter on a doorstep, can be moved to a more suitable indoor environment as need be. Savory herbs in a hanging container can be moved outdoors for a bit of sun, then back into the kitchen. Dwarf mums, flowering inside early, go outside to enhance the balcony, then back inside as fall weather threatens. Some vegetables in containers fit this category too. Cherry tomatoes or dwarf hybrids, even peppers in containers, already in early fruit indoors, continue to yield in outdoor locations. Then, with the advent of frost, they can be moved back inside easily to continue their growth and production.

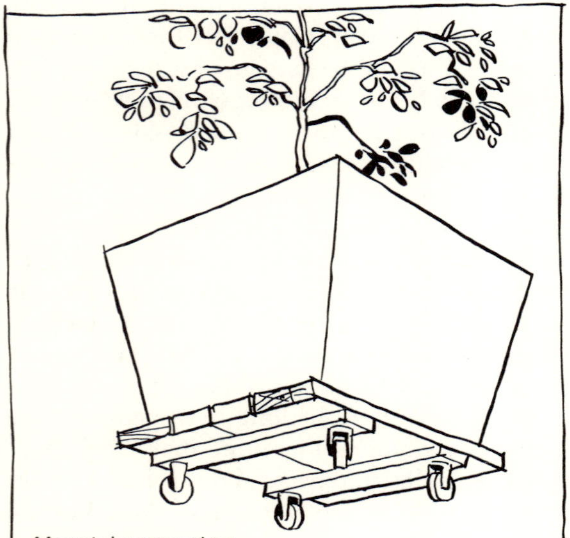

Move tubs on casters.

Or try the old rug trick.

Container Mobility

Mobility is the secret of your success. Small containers you can pick up easily and move about. But when it comes to the larger pots, tubs and boxes you will need easier means.

One thing to remember is that if you plan to move a container, don't water it first, for then you'll have to carry the weight of that excess water. Move them, then water them.

Many containers, particularly planter boxes and tubs, can be equipped with casters, the carpet type are excellent for this purpose. Or you can attach the casters to a small square of heavy plywood and slip this under the container before moving it about. A child's coaster wagon helps in moving reasonably heavy containers. Larger ones have to be handled with skid cloths, skids, rollers or a hand truck.

The skid cloth is an adaptation from the moving-man's bag of tricks. A burlap sack or old throw rug will do. Tip the container up a bit to allow the cloth to be slipped under it. Then, by tugging on the cloth, your container is moved to a new location. It's a surprisingly easy way to pull about heavy containerized plants, but doesn't work going up or down stairs!

A small dolly or hand cart such as professional movers use for heavy boxes is a most useful tool for very heavy containers. These and special trucks for going up and down staircases are available from equipment rental services and are worth their weight in gold when it comes to moving 200-pound containers about.

Another method for very heavy containers is to use several rollers, lengths of steel pipe or heavy clothes-closet wooden rods. You first tip up the box—a shovel is helpful in this—slip a roller or two underneath it, then push the container up on the rollers and move it along, putting rollers from behind up front again.

CHAPTER SIX

Matching Plant to Pot

So far we've discussed the varied uses of container-grown plants and the multitude of benefits derived from having them at your command indoors and out, during any season of the year. The sorts of containers most useful to you are described in the next chapter. Plants you can grow in containers are discussed here.

While your choice is almost unlimited, your selection of plants does depend on certain factors. The use to which the plant is put is more or less obvious. Herbs would be grown for use; flowering bulbs, shrubs, annuals and perennials for their color and beauty. Less evident perhaps is the adaptability of the plants to those special environments where container-grown plants are put to use—as well as the adaptability of the plants, and particularly their root systems, to being grown in a pot.

By special environments I mean, of course, growing plants where they do not grow naturally. In nature virtually all plants are found in some natural environment, to which over millenia they have become adapted. Dependent on their photosynthetic requirements for light and tolerance to heat and cold, dry or wet conditions, plants can be classified as light-loving, shade-tolerant, desert types, tropical rain forest plants, and so forth. Knowing something about the natural growth conditions of plants is quite important in selecting plants for indoor or planned outdoor situations; and in order to know how best to care for them in terms of light, warmth and water needs. The rosters of plants given in the Appendix reflect some of these natural environmental requirements in terms more suited to your use.

While many kinds of plants will grow, or at least survive, under less than optimum conditions of light, heat and moisture, it is best, from the standpoint of the plant's and your own needs (for growing the plants in the first place), to provide the best possible growing conditions. The mobility of containers satisfies this in part, in that you can move your plants into sunnier locations where required or into a shady spot if that is called for, or indoors to escape inclement weather, or to secluded areas for rest periods. You also can augment nature when needs require by watering, providing warmth and artificial light.

Luckily, however, most plants are adaptable to fairly wide ranges of environmental conditions and particularly to those temperatures commonly found indoors. So, if you have selected your plants with consideration as to where they will be grown most of the time, providing them with the best growing conditions is not too difficult a task.

The kind or variety of plant plays a part in its selection, as well as its adaptability to unusual environments, but also does its size, age and condition. The selection of plants to grow in containers then can be defined by their adaptability, usefulness and growth criteria.

These features vary with the kind and variety of plants available, and as plants usually are selected on the basis of their "use," the following comments are based on a plant's usefulness for various situations. But consideration is given also to its adaptability to the various situations: in other words it is fitting the plant to the pot.

Here are some suggestions as to the kinds and sizes of plants best adapted to your needs and uses, set forth in groupings as to those providing color, green background or accent foliage, those whose attributes create vertical or horizontal arrangements, plants with colorful or edible fruit, those useful for food or flavor, or best used in aquatic sites, miniature or contained gardens in glass, for the livingroom, entryway, office room and so forth.

Plants for Form

When selecting plants for a particular use such as providing vertical or horizontal emphasis, or because of need for special attractive or eye-catching shape or form, consideration should be given to how well and how long such a plant will continue to fill that particular need.

Trees and shrubs, strategically located in tubs or boxes are excellent for adding height accents. They can be arranged either singly or in rows. Slow-growing trees and also those with unusual foliage perhaps are best. Norfolk Island pine, Japanese maple, fig trees, conifers, citrus, palms and philodendrons are good examples. The taller-growing shrubs like podocarpus, aralia, azalea, gardenia, holly, oleander, rhododendron and roses also come to mind as useful for height accent. The bamboos fit this category, as does broomcorn.

Vertical effects can be gained through the use of espaliers, standards and topiaries. Many fruit trees can be trained and pruned into the familiar, formal branched espalier shape and used against a wall or fence. Others suited to this are figs, Japanese holly, some Taxus species and camellias. The standards are plants shaped into tree forms. Azaleas, geraniums and fuchsias most commonly are used this way, but camellias and chrysanthemums can be trimmed and trained into reasonably upright forms, too. Usually the plants are staked, lower side branches removed, and later buds and flowers are pinched back (after flowering) to maintain form.

Topiary is an ancient art that involves pruning and shearing upright growing plants into geometric and lifelike forms. Boxwood, holly, privet and some of the cedars or arborvitae are good examples. What is required is dense-growing, much branching foliage to maintain its sheared shape. Synthetic topiary geometric forms may be achieved by placing suitably shaped coarse mesh wire cages around containers of vines mounted on pipes or poles. The vines are allowed to fill and cover the wire mesh shapes. Then these vine-covered shapes are maintained in topiary fashion by judicious pruning. The taller topiaries are useful not only for height accents but become conversation pieces.

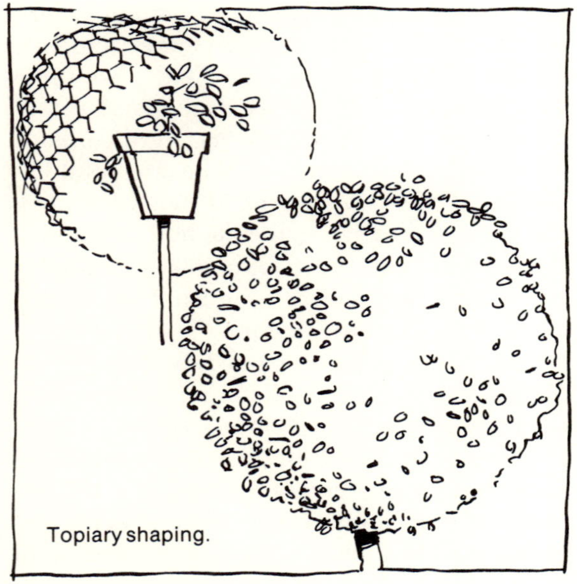

Topiary shaping.

Yet another way to achieve a vertical effect is with hanging baskets, particularly those using the types of plants that will grow down in a cascade of flower or foliage. Almost any of the vines suit this purpose, particularly the ivies, and some other plants such as fuchsia, Emerald Fern, and lobelia. Donkey Tail Sedum will trail down some distance from a hanging container.

Most vines, however, prefer to grow up, and if trained to a suitable trellis will add vertical height accent this way to indoor or outdoor arrangements. They include ivy, morning glory, Moonflower, jasmine, Passion Vine, Star jasmine, and bougainvillea to name only a few. Don't overlook sweetpeas and scarlet runner beans. Pole-type climbing edible beans can be trained to ascend strings or trellis and not only are used for appearance but as a source of food.

Another aspect of form is flat or horizontal in nature. Low-growing bedding plants and ground covers in massed small containers or in larger, low-edged tubs provide the level forms often needed to offset vertical or average height arrangements. Many vines fill this bill if allowed to ramble and spread out, particularly English ivy and geranium ivy. So do many of the sedums, Sandwort and Creeping Thyme. Useful, too, are pachysandra, myrtle, periwinkle, Wintercreeper, and even Creeping Jennie *(Lysimachia nummularia)* or Ground Ivy *(Glecoma hederacea)*, both species of which are weeds! In any case, since your ground cover plants are in containers, there's little chance of their escaping and going wild. Some care needs to be taken, unless the arrangement is to be somewhat permanent, to prevent those that can from rooting elsewhere along the vines.

Another way of using plants in a particular location to enhance form and dimension is in eyelevel hanging baskets of cascading vines or flowers, but which do not trail down to a length for height accent. Mesh baskets are the usual containers, although any other sort that will provide good drainage and can be hung from small chains, wire, or rope, will do. Even a pot imbedded in a basket of spaghnum or Spanish moss is good. Plants suited to this eye-catching use are Asparagus fern, campanula, fuchsia, "Pendula" tuberous begonia, lantana and, of course, the cascade varieties of petunia and ivy-leaved geraniums.

One additional concept of plants selected for form includes bonsai arrangements and specimen plants, particularly those with unique foliar characteristics and used as entryway sentinals, lobby and waiting room singles.

Bonsai is a separate art all in itself and not within the province of this book, although many bonsai arrangements are based on a container of some sort. Dwarf conifers are particularly useful, as are some of the fruiting deciduous and citrus trees. Others well suited to this technique are noted on Page 61.

Here form and scale are vitally important, and these features should be kept well in mind when selecting plants for bonsai arrangements. The container is equally important and must complement the size and shape of the plant. Bonsai technique involves root pruning to maintain the plant in small size, using very little or no fertilizer to reduce growth rate, trimming and training branches in place with wire, and pruning to a shape that is aesthetically pleasing.

While those suggested on Page 62 are the

most widely used plants for bonsai, some are better adapted than others. In fact, almost any interestingly-shaped perennial that can be trained and sculptured into shape serves well. Try it, and if a plant doesn't work out it can be repotted for other use.

Plants for Foliage

Here your consideration in selecting a plant would be on the nature of its greenery; the size, shape arrangement, even the color of its leaves. Perhaps all these factors are important. Taken or seen together, this is what gives visual impact. Picture the masses of solid green, delicate lacelike textures, interesting shape and form of the leaves of the castor bean or rubber tree, palm or Japanese maple. All these illustrate differing ways of using foliage. The descriptive terms of graceful, feathery, palmlike, fernlike, roundleaf, ornamental, almost name the plants themselves.

Many trees and shrubs used for accent or colorful flower or fruit also are used simply because of their attractive foliage. All the conifers and palms would fit this situation, and many of the deciduous trees and shrubs as well, not to neglect azaleas, rhododendrons and holly. Some examples are bamboo, beech, cedar, fir, Mediterranean fan palm, Parlor and Sentry palms, Japanese aralia and maple, Sweet Bay, New Zealand flax, philodendron, podocarpus and the colorful bromeliads. All of the vines can be considered foliage plants, as can groundcover plants and some of the annuals such as variegated begonia and nasturtium, also the succulents and cacti, some of the herbs like chives and thyme, and even among vegetables pepper, purple cabbage, squash or cucumber vines.

If foliage of some particular form, shape or color is your goal, make your plant choice on this basis, and if it turns out also to be a flowering or fruiting plant enjoy that as a bonus.

Plants for Flowers

The first to come to mind here are the flowering annuals and perennials, and perhaps shrubs. If your search for a plant to match your pot is based on colorful flowers, you need to keep in mind not only the desired color but the extent and duration of the blooming period and also any special environmental conditions that may be required to bring your plant to flower.

Most annuals provide temporary color only and for a season only, but are excellent to satisfy short-range color needs. If started ahead from seed under lights (see page 17) they provide almost instant color to your room or patio.

With proper growth needs attended to, they will continue to bloom over a fairly long period. Also, because they are easy to replace when in containers, a continuous show of successive blooms of different plants can be achieved.

Many annuals can be adapted to container life and among the better suited are ageratum, Balsam, browallia, calendula, candytuft, Cockscomb, coleus, lobelia, nasturtium, Nemesia, primrose, sweetpea and verbena. Do not overlook petunias, pansies, marigolds and zinnias, for these common garden plants do very well in containers.

One way to handle numbers of annuals for massed effects is to keep them right in their peat pots, which are inserted in other larger containers, tubs or boxes. They easily can be moved about and replaced as conditions calls for it.

Perennials can furnish continual season-long color. Some flower the year-round if growth conditions remain optimum, but in more inclement areas this may mean transport inside for a period. However, many perennials furnish only a brief flash of color, and all need a rest period, which occurs naturally after their period of maximum bloom. So if they are spring or summer-flowering, they would rest over the winter. Those that bloom in fall or winter, rest during the spring and summer. Often during this rest period they can continue to serve as foliage plants and height accents. Many, though, should be cut back after flowering and retired for a rest in an out-of-the-way spot. Winter "resters," if not cold hardy, can be kept in a sheltered garage or semi-lighted basement. Summer resters can be sunk, pot and all, in a secluded flower bed. Some such as the flowering deciduous fruits, the apple for example, require a fairly cold resting period to ensure that next year's buds will bloom.

Some excellent flowering perennials are acanthus (Grecian Urn Plant), agapanthus, aster, begonia, Bell-flower, Bird of Paradise, carnation, Christmas rose, chrysanthemum, cineraria, geranium, Lantana, Marguerite and Plantain lily.

We also should include in the container-grown flowering plants the flowering shrubs and trees, roses, and the bulbs. Azaleas perhaps are the most popular of the flowering shrubs and are useful as foliage plants when not blooming. Dwarf flowering fruit trees such as Japanese cherry, the flowering apricots, crabs, plums and quinces often are noted, but dwarf varieties of peaches and apples also would qualify and would furnish edible fruit as well as a brief but colorful and sweet-scented bloom.

Other flowering trees adaptable to container

gardening are crape myrtle, dogwood and magnolia. Flowering shrubs include Bougainvillea, camellia, daphne, fuchsia, gardenia, hibiscus, Cape leadwort and Carissa, hydrangea, Lily-of-the-Valley shrub, oleander *(Nerium),* rhododendron, shrimp plant, viburnum and one fittingly named Yesterday-Today-and-Tomorrow, whose flowers change from purple to violet to white on successive days.

Roses of all sorts, particularly climbers, floribundas and hybrid teas, do exceedingly well in containers. The new miniature roses are excellent indoor-outdoor plants and succeed very well under lights.

Bulbs in containers, their culture described in Chapter III, offer a succession of instant bloom and easily can be taken on and off stage. Including those propagated from tubers, corms and the like, this group of flowering plants, easily adaptable to container culture indoors or out, would include achimenes, amaryllis, Blood lily, canna, crocus, cyclamen, freesia, hyacinth, Kafir lily, true lilies, nerine, ranunculus, and scilla. But don't forget daffodils or tulips, perhaps the most favored of all the bulbs.

For color we also should consider the fruit and berry plants. Deciduous and citrus fruits have been noted before. New dwarf varieties of citrus fruit, excellently suited to container culture, now include kumquats, lemons, limes, oranges, tangarines and tangelos. Cotoneaster has red berries as do Chinese, Japanese and the traditional English holly, although these are not edible. Other interesting fruit trees available for containers include guava, loquat, mahonia (the blue-black fruit of Oregon grape, *Mahonia aquifolium,* are also edible), plus pomegranate and pyracantha.

Some of the vegetables well could be included in this category too: melons, gourds, Indian corn, peppers, eggplant and especially the midget tomatoes. Several seed catalogs now carry special listings of such vegetables.

Plants for Food and Flavor

Plants in containers can be useful as well as ornamental. You will note several references in this book to salad or herb gardens that grow in your own kitchen. While most of the edible and fruiting plants can be container-grown at least part of the time, some are better suited than others. Plants which grow rampantly and overlarge or which must grow in great profusion to obtain satisfactory yields generally would not be selected for container growth. Here are examples, and some exceptions, of vegetables:

Pumpkin and squash vines take up so great a space they usually would not be recommended for container growth. So do cucumbers, strawberries, raspberries, blackberries, and grapes. Tomatoes, cabbages, sweetcorn, sweet and white potatoes, and herbs such as bay, dill and fennel require a lot of space, too. Usually quite a number of plants are required (usually covering hundreds of square feet) for pop, sweet, or Indian corn, rice, wheat, soybeans, broomcorn and sorghum in order to obtain any sort of yield. However, every one of these plants *can* be and has been adapted to container growth for one purpose or another.

The most common container vegetable is the tomato. With the advent of hybrid, midget varieties such as "Small Fry," "Patio Hybrid," "Tiny Tim," and others, tomato plants yielding excellent small to medium-sized fruits on plants not more than a foot or so high now can be cultured conveniently in containers. Such a small plant covered with appetizing red fruit is decorative, as well as yielding a fine source of salad ingredient.

Other possibilities in the roster of container-grown vegetables, fruits and herbs would be a tub of squash also grown for its attractive leaves and flowers, a crockery pot of cascading cucumbers, strawberries (particularly everbearing) in jardeniers, crocks of raspberry or blackberry plants trellised in semi-espalier fashion or a tubbed grape trained vertically for height.

Even pots of cabbage (especially purple cabbage), kohlrabi, peppers and eggplant are useful as decoration and eventually food yield. Peppers are particularly colorful, since varieties are available with red, yellow or golden fruit and they, as the tomatoes, now come in midget, hybrid forms. An eggplant's dark purple fruit is a most colorful as well as an unusual container-grown plant. Spanish peanut, with its yellow flowers, is unusual, too. You never would be able to grow enough potatoes in a container garden to feed a family, but a single plant in the classroom is quite an educational experience.

Even corn and the small grain seed crops have a place in container or light garden culture. A tub of sweet or pop corn can add height and yield at least a taste of edibles. Rice also can be cultured for vertical accent in a pool arrangement, while wheat sometimes is sprouted in quantity in flat, traylike containers to yield highly nutritive green growth, as other grains also are sprouted for food.

I noted bay, fennel and dill as herbs not usually recommended for indoor culture. But bay is a small tree and is admirably suited to a tubbed existence on the patio, being quite amenable to shearing and shaping. Dill and fennel usually grow rampant and tall, but new, short varieties are available. If you grow them for their leaves, not seeds, they can be kept small enough to include in a windowsill garden.

So you see, the exceptions can be the rule, when you read that virtually all kinds of plants can be grown in containers. This is pretty much true, and it is just as true of plants grown for food and flavor. Some obviously are grown just for appearance or to have something unique—like a pineapple or banana plant—and sometimes unusual vegetables such as a sweet potato vine or a scarlet runner bean are grown as a demonstration or experiment in the classroom, or perhaps just for fun. All can be and are grown in containers.

Of all the vegetables, however, tomatoes perhaps are the easiest and most rewarding. The midget varieties, as noted, are best suited to container culture. Perhaps as easy to handle, though, are others of the salad group: lettuce, radishes and carrots. Leaf lettuce makes an attractive pot plant, grows rapidly and yields well. "Buttercrunch," "Salad Bowl," "Oakleaf," and the Bibb types are best. Spinach, too, and Swiss chard belong in this group, although the latter usually is cooked. When your containers will be outdoors and it gets warm, Swiss chard will grow well while spinach will not. But the New Zealand types of spinach and tampala, a spinach-like oriental vegetable, prefer hot weather. Radishes are the quickest of all to grow and do well under lights with not too much heat but lots of water. Most popular are the round red types such as "Cherry Belle" and "Champion."

In the salad group we also could place carrots and onions. Tender baby carrots, only a few inches long and as thick as your finger, make a tasty salad addition. Try the new midget varieties: "Short-n-sweet," "Tendersweet" and "Tinytot." For onions the quickest results can be had by growing them from "sets" or young plants available either from your catalog or local garden supply house. White bunching types are preferred to yellow, or to those grown for their large bulbs. Young plants are ready to eat as scallions or greentails in about three weeks, sets in about five weeks.

Cucumbers also belong in the salad group. Started from seed in peat pots, they then are transplanted to their permanent containers. The vines are allowed to cascade or they can be trained on a trellis. Select one of the new, F_1 disease-resistant "slicer" hybrids that will mature small to medium fruit in seven to eight weeks. Chives, parsley and other herbs are noted later under herbs.

The other salad vegetables are the peppers, their very colorful fruit green at first, then turning yellow, golden or bright red. They are grown easily in containers, require fairly warm conditions and once mature will continue to bear fruit for quite some time. If your container-grown peppers are on the patio they can be brought indoors come frost and if kept pruned back will continue to yield all winter. Many varieties are available. Select one of the new hybrid Wonder or Canape varieties. A few midget varieties now are listed in the catalogs, and you might like to try these, or perhaps even a "hot" pepper or two for variety.

Eggplants are fun for container growing, are attractive, and also will furnish you with edible fruit. They require as much or more heat than peppers and tomatoes, so they do best on a sunny patio. Grow them, one plant to a container, from purchased seedlings, or raise them yourself from seed under lights. If you do this give them a two-month headstart. Pinch the terminal stems and blossoms back so that only four to six fruits will form per plant. The new dwarf varieties such as "Morden Midget" are excellent, as are other standard varieties having smaller plants.

Beans and beets grow well in containers, too. Sow quite a few beet seeds in a rather large container and then thin out the extra plants, using them for salads or as cooked greens. Harvest the rest when an inch or inch-and-a-half in diameter. They taste much better when a little on the young side. The standby favorite is "Detroit Dark Red," although "Ruby Queen" has shorter tops that are quite attractive as foliage plants, and the beets themselves are tender and delicious.

Beans—that is the bush-type varieties of both wax and green as well as limas—can be grown. You'll need a group or row of at least ten containers to furnish the average family with several pickings. They do best on the patio, and should be sown directly in the containers. The small "butter"-types of limas are very good, as are the "tender" varieties of snap beans, "Tenderpod" and "Tenderette." For interesting variety, particularly where height or some vertical screening is needed, trellised beans of the pole variety are a good choice. The so-called horticultural beans, both bush and pole, have very colorful pods splashed with crimson and are tasty when cooked.

Squashes do well, too, as patio plants when grown in moderately large tubs or other containers. Select from varieties having compact, bushy vines and keep them in a sunny location. They require quite a bit of water, too. The summer squashes best for container culture include the scallop or patty-pan types, and a new midget variety called "St. Pat." The Gold Nugget's bright yellow flowers, golden fruit, and large green leaves make it a good choice, too.

Zucchini also does well, particularly the cocozelle bush types. Harvest zucchini squash when the fruits still are small, about the size of a small cucumber. The winter squash—the butternuts, acorn and hubbards—take from 80 to over 100 days to mature, but if you have the space and the time, they make rewarding and unusual container plants. Many are grown readily in containers both indoors and out.

Herbs

Many herbs are grown readily in containers, both indoors and out. Common or Italian basil, an annual and easily started from seed, is suited best to container culture. Keep the centers or terminal shoots pinched back to prevent it from going to seed and it will continue to furnish you with spicy leaves for spaghetti and tomato dishes and salads.

Bay's other name is Laurel *(Laurus nobilis)* and unattended to, it will grow into a small tree. However, by judicious pruning and by virtue of its relatively slow growth, you can manage one easily in a large container. Its leaves are used to flavor a number of dishes including spaghetti, soups, and curries.

Chives are among the most favored of all container-grown herbs. Cut fine, it will season soup, salad, cream or cottage cheese, omelets, potatoes and a number of other dishes. Cutting the plant only makes it grow better. It is easier to start a pot of chives from divisions of an older plant, or young plants from the greenhouse. It takes two years for chives to mature from seed. Plan on more than one pot.

Dill and fennel, as noted earlier, are best grown for their leaves. If needed for seasoning the seed of either is better secured in grocery stores. Grown and harvested continually for their leaves, neither plant will outgrow its bounds. Dill and fennel leaves, known as "weed," both are used in fish dishes and with cream or cottage cheese spreads, although dill perhaps is best known for the use of its flowers and seeds in pickles.

Marjoram is grown easily from seed in a container, seldom exceeding six inches in height. Leaves are taken from the top of the plant—the terminals—to prevent it from blossoming and thus shortening its useful life span. It is used in a number of dishes including egg and sausage and with pork, duck and lamb.

Mints are used largely in teas but also in jellies to serve with lamb dishes. There are several useful species including apple mint which grows a foot or so tall. Orange mint seldom exceeds six inches in height and, as with apple mint, its name suggests its flavor. Pineapple mint has variegated leaves that cascade strikingly when grown in a hanging container. The best known of this group of plants are peppermint and spearmint. Varieties are available that adapt well to container growth and seldom exceed a foot in height.

Nasturtium is unique among the herbs in that it is often grown as a decorative flower but its flowers are edible and can be served stuffed for canapes, the leaves used to season salads and sandwiches. Several forms are available—short compact varieties for small pots, climbing varieties to train on a trellis. All are started quite easily from seed.

Oregano, a member of the mint family, grows from six to ten inches tall. Most common are the sweet, mint-smelling hairy-leafed type and the small, smooth-leafed variety. It is used mostly in spaghetti or on pizza, but try a bit sprinkled with lemon juice on pan-fried chicken livers. Picking top leaves will keep the plants bushy and productive.

Parsley, the king of the herbs, is used in so many ways that they don't need mentioning. This is a biennial, started from seed in peat pots and kept in a cool, semi-shady spot until well developed. The young plants then are transferred to their containers where they will grow to six or eight inches in height and yield for many months. Pick only the outer leaves or sprigs as the plant develops new growth at its center.

Rosemary comes in prostrate, creeping and upright forms. It can be started from seed but takes several months to develop, so perhaps you will be better off to buy young plants. They grow quite slowly indoors and won't need much trimming to keep within bounds. Rosemary leaves are used to season jelly, lamb and chicken.

Sage, one of the better-known herbs, is grown easily from seed and is quite adaptable to container culture since it seldom exceeds a foot in height. It needs quite a bit of light and sweet soil for best growth. Its best-known uses are for seasoning sausage and poultry stuffing, although in Vermont it long has been used to season cheese.

Savory is grown both indoors and out. The annual summer savory is preferred for seasoning stuffings and with snap beans. It grows slowly to a height of six to eight inches. Winter savory is a perennial and does very well on the kitchen windowsill. Both require quite a lot of sunshine for good growth.

Tarragon, difficult but not impossible to grow from seed, is best started from root cuttings or divisions. It doesn't need much sun or water, but should not be allowed to dry out completely. Leaves from the top or sides of the plant are used to flavor vinegar, fish and meat dishes.

Thyme is a short plant and even the upright-growing varieties seldom exceed ten inches in height. Both creeping and upright types are started easily from seed but grow slowly and require considerable light or sunshine. Seedling plants often are available from greenhouses or nurseries. Thyme is used to season soups and chowders, and with pork and lamb.

Plants for Fruit

Many fruiting trees and shrubs, deciduous and citrus and others, are quite adaptable to growing in containers. Many do well outdoors and can be brought in for short periods of time, while others survive most happily with an indoor-outdoor existence. A few do best indoors all year round. Fortunately dwarf forms of most fruit varieties now are available.

Plants for fruit that can be grown in containers include strawberries, especially the ever-bearing varieties which look well in many-pocketed terra cotta strawberry jar. Other small fruit would include grape, elderberries, currants, gooseberries, blueberries, blackberries, raspberries, boysenberries and other "bramble" fruit. I have never tried these in containers but see no reason why you couldn't.

The dwarf citrus trees include those that have been budded on dwarfing rootstock (and which can be kept under three feet tall when in containers), and the natural dwarfs. The latter include the "Meyer" lemon; the "Otahiete" orange (whose fruit looks more like a lime); the sour but beautifully deep-orange colored "Chinotto" orange; kumquats such as "Nagami" and calamondins.

Just about any citrus variety can be budded on dwarfing rootstock and so can be very suitable as container plants. These would include the sweet Valencia, blood, and navel oranges, tangerines (mandarin oranges), limes, lemons and even grapefruit.

Citrus do best indoors but can stand and even like, short vacations on a sunny patio. You can raise your own citrus, too, from seed saved from breakfast fruit, but just for fun, since such a plant won't set fruit. Plant them in potting soil mix and place under the lights.

Deciduous core and stone fruit trees also are available in dwarf form. They are kept outdoors most of the time and are brought indoors only on special occasions. If kept indoors during their entire flowering period, they will need to be hand-pollinated if you want fruit. These core fruits include apples, crabapples, pears and quinces. The stone fruits are apricots, cherries, nectarines (of which some varieties such as "Golden Prolific" and "Nectarina" are natural dwarfs), peaches and plums. Other fruit trees now available in dwarf form include oriental and American paw-paw, persimmon, pomegranate, fig and mulberry.

You also can grow avocados, dates, mangoes and papayas, even coconuts from seed, but only for fun and for a while. They soon outgrow their container and seldom will yield fruit.

Plants for Fountains and Pools

Container-grown plants are used in two different ways: arranged around the edges of the pond, pool or fountain or submerged in the body of water itself. Many sorts of plants in containers fulfill the first purpose. For the latter, of course, only those that thrive in an aquatic environment will do.

Basically there are two types of aquatic plants—those that grow up to and on the surface of the water and those that emerge and grow to some height above the water level. Among those that do well, rooted in submerged containers, are water lilies. Their rounded leaves float on the surface, and they have quite showy, star-like white or rounded yellow flowers that also float or stand above the water on short stalks. Among the best of the dwarf types are *Laydekeri* and *Phgmae helvola*. Select hardy types that can stay in the pool all winter. Among the tropical varieties that must be stored and replanted annually where winters are cold, White Star is one excellent variety.

The lotus, with its very large fragrant and showy flowers, is another rooted, submerged aquatic plant. It grows best in round containers, such as clay ¾ pots, with lots of sunshine, and it must be stored during winter. Water snowflakes have small white flowers on short stalks above the water, with small rounded leaves like water lilies, and like the water lilies, they like lots of sunshine. Water hyacinths, however, will do well in a somewhat shady location.

Of the upright growing plants, perhaps the most useful for vertical accent is the Japanese iris. A pot half-submerged towards the back of the pool will add a most interesting accent of long, narrow pointed leaves and colorful blooms. Also quite useful are clumps of horsetail rush, dwarf papyrus and cattails. All do best in porous containers half-submerged in the water.

Many different sizes and varieties of pool plants are available. A careful selection and planting in proper proportions will enhance any aquatic layout greatly.

Plants for Porch or Patio

Some plants especially fit the indoor-outdoor

category since they thrive indoors as well as out on porch or patio. Geraniums are the standby. They come in red, white, or pink and in muted but colorful blooms. Their pretty foliage often is fancy and variegated. Some are sweet-scented and some trailing, making them excellent for hanging baskets. Other good choices, including those mentioned earlier, are the ivies, philodendron, Sentry palm, Parlor palm, coffee trees, Fiddleleaf fig, Threadleaf False aralia and sansevieria. Any of the hanging or basket plants can come indoors, as well as many flowering shrubs and fruit trees. The citrus group is particularly well adapted to this shifting of climates, and there are many others.

Plants for the Parlor

With proper care and easily-moved containers, most plants can travel in and out and need not remain housebound. However, there still remain a few plants that could be catalogued strictly as house plants, though even then some could go outdoors for a short periods of time.

These include the usual ferns, philodendrons, palms, geraniums, begonias, African violets and other gesneriads, and the orchids. These latter require special attention as to light and moisture, need high intensity light, high humidity. Their containers, usually porous clay pots, are embedded in a nest of sphagnum moss. Their planting mix usually includes shredded osmunda fiber. Orchids include the common florist type or cattleya and also the elegant ladyslipper orchid or cypripedium, the sprays of oncidium and the Lily-of-the-valley and striped orchid or odontoglossum.

Other houseplants include anthurium, Shrimp Plant, the bromeliads, cineraria, cyclamen, fuchsia, heliotrope, impatiens, Lantana, and the Apostle plant. Among the foliage plants, I also would include Asparagus fern, Aglaonema, aspidistra, coleus, dracaena, dumbcane, Kangaroo vine, mimosa, Prayer plant, pandanus (screw pine), peperomia, Piggyback plant (or pick-a-back), rubber plant, sansevieria, tradescantia and zebrina.

Plants for Miniature Gardens

Many kinds of plants are tiny by nature, and others can be trained to grow so. There are, also, the many new midget and dwarf hybrid varieties. All can be used to create miniature garden arrangements in a tiny, eyepleasing setting, such as in a tray or dish. In Chapter XII on gardening-in-glass we list a number of plants adapted to such gardening.

Dish gardens—and this term includes a whole variety of shallow containers—usually are assembled with plants all of a similar environmental liking, and so the most common arrangements often are referred to as desert, tropical or woodland. The useful desert plants include many of the tiny cacti and succulents such as *Andromischus maculatus, Astrophytum myriostigma, Crassula argentea* (jade plant), *Crassula schmidti, Crassula teres, Gasteria lilliputana, Haworthia tessellata, Portulaca,* and any of the tiny sedums. An arrangement of these would be kept fairly dry and in lots of sunshine.

The tropical dish garden might well include miniature orchids and ferns, some creeping ground cover and specimens of *Allophytum Mexicanum, Carissa grandiflora,* dwarf Euonymous, *Kalanchoe pumila, Leptotes bicolor, Masdevallia simula, Oncidium pulcheleum, Tillandsia ionanthe* or miniature Rex begonias.

Several native plants that can go into miniature woodland gardens are in Chapter XII, as potential components for terrariums and bottle gardens. Such arrangements will do best in semi-shade, only moderate light and at 60° to 70°F. Useful in such arrangements, too, and often available from florists, are miniature azaleas such as "Gumpo," the tiny begonias *(Begonia aridicaulis, B. boweri, B. dragei),* also the small bromeliads *(Crypthanthus acaulis* and C. *zonatus), Echeveria glauca,* the miniature ivy *Hedera helix* var. Glacier, *Kalanchoe* var. "Tom Thumb," *Neanthe bella,* creeping *Pilea depressa,* strawberry geranium *(Saxifraga sarmentosa), Sempervivum arachnoideum,* a miniature gloxinia *Sinningia pusilla,* tiny form of *Oxalis,* and others, including the grass-like *Acorus gramineus pusillus,* the rainbow flowers (Achimenes), a fern—*Adiantum hispidulum, Althernanthera versicolor,* African violets, *Calanthea argyraea,* the silver-spined bromeliad *Dyckia fosteriana,* grape ivy *Cissus striata,* the jewel orchid *Erythrodes nobilis argyrocentrum,* miniature Peperomias, and a greenleafed creeper, *Selanginella kraussiana browni.*

Plants for Bonsai

One very attractive and currently popular form of miniature gardens, although they may involve just one plant in a container, are those referred to as bonsai. These plants are not always naturally tiny but frequently are dwarfed by careful root and top pruning and are trained into graceful forms. There are several good references on the art of bonsai culture, and this indeed can be classified as a unique form of container plant culture. Many small woody plants that adapt to bonsai growth include tiny woody begonias, *Camellia sasanqua,* small Cotoneasters, dwarf types of Chamaecyparis, miniature Ficus, *Ginkgo biloba;* many of the conifers such as Sargent juniper, Norway, mugo, and Japanese pine, and Norway spruce. Also good are Kurume and other dwarf azaleas, trident and Japanese maple, tiny flowering quinces, dwarf pomegranate, dwarf false cyprus, and *Zealkova serrata.*

CHAPTER SEVEN

Matching Pot to Plant

Volumes are available extolling the beauties and virtues of the many containers available in which to grow a plant. But little has been offered on the hows and whys of container selection and use. How a plant will grow best in what container — in fact how a plant grows at all in the confines of a pot, jar or tub—these important questions not often are answered. Some basic knowledge about how a plant grows and about its proper growth environment are essential to have in mind when selecting a container. The container itself is an important and often critical component of this system of gardening.

Just about any sort of recepticle can be used or adapted to container gardening, provided that it meets certain criteria. These are: how well it suits the plants' needs, and, of course, yours. The latter is a much more flexible yardstick, for how desirable you consider a container depends largely on its aesthetic qualities. Other considerations are cost, durability and transportability. Again these are personal options.

How well a container suits the needs of the plant is a most important consideration, and here I am talking about how well the container will function in helping provide the proper plant growing—especially root-growing—medium or environment. The container can influence the effects of any of its contents—soil, organic matter, beneficial microorganisms, nutrients, air and water. Proper container choice will ensure that this influence is positive and contributes to the well-being of the plant.

Containers should have a hole or holes for drainage of surplus water so that the proper balance of air can re-enter the soil after a watering, some air being essential to good root functioning. The air for the roots of aquatic plants in submerged containers is furnished by the leaves and stems of the plants themselves.

Porous materials are best for the same reason. As moisture evaporates from their outer surfaces, it tends to keep the soil-medium and roots cool and the air near the plant more humid. However, as porous clay or terra cotta containers tend to dry out quite rapidly in warm weather, this could be a disadvantage. You can prevent this by burying the pots in a nest of moist sphagnum, peat moss, vermiculite or even sawdust inside another container or jardiniere.

Non-porous containers, such as those of glazed terra cotta, stoneware, glass, ceramic and plastic, will not need watering so frequently. Drain holes in them are absolutely necessary and they must be kept functioning. Salt encrustations of fertilizer dried on the outside of these pots does not occur, as sometimes it does with porous pots. Since moisture is not evaporating from their outer surfaces, non-porous containers do not keep the pots' contents as cool nor the air above the pot as humid.

Wooden containers have a high insulating value in that wood resists heat penetration and the pots' contents stay relatively cool even when the containers are in direct sunshine.

I have been emphasizing a container's ability to retard heat. This is fairly important, particularly for container-grown plants on a patio in sunny climates. Roots generally function better at temperatures several degrees cooler than those best for the above-ground portion of the plant.

Wood is used generally in the construction of larger containers because of the saving in weight. But it is subject to decay, and unless constructed of decay-resistant wood or treated with some preservative it will not last long. Cedar, cypress, black locust, osage orange are favored wood, and sometimes chestnut, though this is becoming scarce. The most popular and available wood is redwood. Decay-prone woods can be treated with preservative or asphalt compounds, but be sure you use a type recommended, for some are harmful to the plant roots.

Wooden boxes, tubs, kegs and soytubs frequently are constructed of vertical staves held together with metal or wooden bands or hoops. These should not rest with all the weight on the stave ends but should be propped up on cleats so that the weight of the container and its contents rest on the cleats with the lower ends of the staves clearing the ground, so to speak. In fact, any wooden container should have slots or holes in the bottom for drainage, and the bottom kept somewhat dry by being raised an inch or so on cleats, bricks or blocks of wood.

Rustproof sheetmetal or polyethylene plastic liners are useful to line the insides of wooden containers—especially uniquely shaped ones such as old butterchurns—to preserve the wood.

Another consideration is container size. This is roughly dependent on the size of the root mass that you wish the container ultimately to accommodate. Overly-small containers will cause restricted root masses and slow down top growth. Some restriction is useful in keeping miniature plants tiny, especially in bonsai culture, but too much restriction leads to rootbound or potbound conditions, and unhealthy plants. Overly-large containers may favor root growth over top growth or just be wasteful of space and soil and create unnecessarily heavy containers.

Containers come in all sizes, literally, from thimble-sized mini-pots or ounce-sized bottles for a bottle garden to gigantic redwood tubs holding several tons of soil and full-sized palm trees. The choice is yours.

So one can see that a container's functions are in relation to its size and the substance of which it is constructed. This affects whatever is inside the container—the air, water and nutrients for the roots, an amicable soil temperature—and of course the anchorage or support for the plant.

Types of Containers

Containers usually are selected first on the basis of size and second on material, although you might be looking for a certain type and color of container for a specific use, such as an earthy-brown, unglazed terra cotta dish for a bonzai arrangement, or a redwood tub for a favorite azalea.

Your choice, too, might be influenced by where the container will go , such as a glazed urn for an entranceway fern, a tub for a water lily, or a special candy jar for a bottle garden gift.

For convenience sake the following containers are grouped according to types—that is, what they're made of—and the sizes and shapes available.

Conventional unglazed clay or terra cotta pots: These come in top-diameter sizes from two to sixteen inches and in colors ranging from chocolate brown through brick red, yellow tan to a beige white. They are about as tall as they are wide and tend somewhat to tip over. Being unglazed they will dry out rapidly.

Standard unglazed 3/4 pots also are called fern or azalea pots. As noted earlier, they get their name from the fact that they are three quarters as high as they are wide. Sizes range from four to fourteen inches in diameter.

Bulb pans, sometimes called seed pans, resemble deep saucers but these unglazed terra cotta containers come with a drain hole in the bottom. They range from six to twelve inches in diameter and usually are one half as deep as wide.

Unglazed terra cotta pot forms also include those that curve gracefully inwards toward the top; the Venetian pots that come seven and one

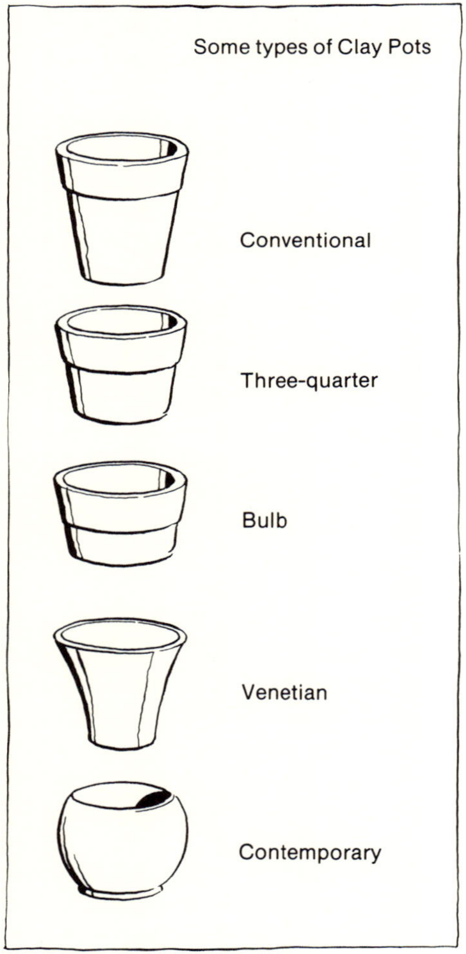

half to twenty inches wide and have sides that slope outwards giving the top a slight flare; Spanish pots that come in eight- to twelve-inch diameters. Spanish fern pots are a bit more squatty and come up to eighteen inches in width. Also available are unglazed terra cotta bowls, urns, strawberry jars and other uniquely-shaped containers including those with classic bonsai lines.

Glazed terra cotta containers come in all the above shapes, sizes and colors, and even more. Some are fluted, others ribbed or ringed. Colors include almost all the hues of the rainbow, and are embedded in the glaze or painted on.

Concrete tubs, urns and other large containers, either conventional, lightweight or combined with plastic such as "plasticrete" can be found in florist and garden supply centers. Almost all are made by pouring the concrete mix into a mold, so the range of shapes and sizes is almost unlimited. The conventional concrete containers are quite heavy.

Molded aggregate, stoneware and porcelain containers also come in many interesting forms, some quite modernistic. A new synthetic ceramic material, Ciorgoda Synceram, is being used to cast cylindrical, square, rectangular and contemporary shaped planters.

Ready-made wooden tubs and planters usually are constructed of redwood, although I've seen some made of treated and painted pine. Hexagonal tubs come in eight- to twenty-four-inch diameters. Round ones made of vertical staves held together with wire or strapmetal bands or hoops usually are from six to eighteen inches in diameter. Square redwood tubs come in ten by ten-inch or twelve by twelve-inch sizes. Flared planters, much wider at the top than the bottom, come in square or rectangular shapes nine to ten inches wide, seven inches deep and twenty-four to forty-eight inches long. Square-sided redwood windowboxes also are available commercially as are redwood hanging tubs, baskets and novelty planter items such as wheelbarrows and donkey carts.

Every day we see more of the synthetics and plastics used in ready-made containers. They range from reinforced plexiglass tubs to styrofoam, foam polyethylene and both rigid and flexible plastic pots. Fluted, thin-walled plastic pots are replacing the conventional metal cans for containerizing nursery stock. Some plastic flower pots come equipped with attached or detachable saucers, even with sponge or wick inserts for "automatic" watering devices. Plastic hanging baskets and pots and half-round pots that hang against the wall are available, too.

Glass vessels usually are not mentioned as containers but they, too, have a place. Unless equipped with drain holes (which are difficult

Some types of Tubs

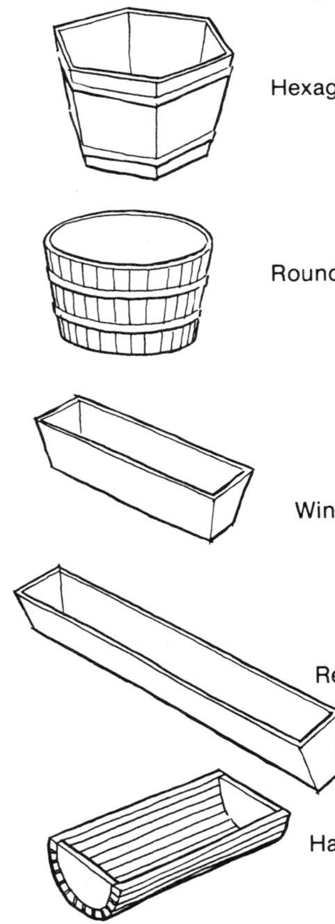

but not impossible to drill) they serve better as outer containers than as something in which a plant is grown. Of course glass containers, ranging from aquariums and fish bowls through all sorts of jars, cannisters and bottles, are quite useful for terrariums, berry bowls and gardens-in-a-bottle. The more unusual the glass vessel, the more highly prized the container.

Jardinieres and cachepots have been mentioned before. These fancy containers usually have no drain holes and are used as attractive outer covers for mundane or soiled pots. All sorts are available commercially, many made of sheet brass although some of plastic. You can convert any jardiniere or cachepot made of suitable material into a useable plant container by putting drainholes in the bottom.

One other sort of container is made from pressed or molded paper, a fiber such as sugarcane residue (bagasse), sphagnum, peat moss, or clay-peatmoss and fiber-peatmoss mixtures. Also asphalted strips or open-ended squares are assembled into bottomless "plant band" containers. Except for those that are waterproofed or asphalted all are biodegradable, gradually will break down and disintegrate when wet and filled with soil. This is a distinct advantage over the clay pots, when it comes to handling transplants. The seedlings or rooted cuttings can be left right in these peat pot containers and transplanted, container and all, to their permanent container or garden home. These decomposable pots come in sizes from one to four inches in diameter and are round or square in cross-section.

Container Care

Not much is required usually in the way of container maintenance except possibly a subsequent treatment with stain, wax or preservative for those made of wood. Terra cotta and concrete or ceramic containers sometimes need washing. Unglazed pots in particular become stained with soil, algae or fertilizer salts. A simple soaking usually is all that's required, perhaps with one or two changes of the water. Stubborn stains may take a while to soak off, and only infrequently you will need to resort to the scrub-brush. Glazed and similarly hard-surfaced containers can be cleaned up in place with a mild detergent solution.

You can make your terra cotta containers look like new after their water bath by baking them in a moderately hot oven for a few hours. This treatment also will kill off any remaining algae or other unwanted microorganisms.

New pots or cleaned old ones often are quite dry, especially those of unglazed terra cotta.

They should be soaked in water for several hours prior to being used, so that the porous clay will not absorb the water needed by the plant.

Homemade Containers

With so many ready-made containers available it would seem hardly necessary to suggest making them. However, the only way to get a certain shaped and sized container may be to build your own. Often many household objects can be converted into economical containers, especially for transplants and for temporary quarters until the proper container is available. Also many an article whose original use was for something else can be converted into interesting or unique containers, jardinieres or cachepots to hold a less-elegant pot.

Wood is the most popular building material for do-it-yourself containers. Select a rot-proof wood such as redwood, cedar or cyprus. If this is not available, treat the interior with a wood preservative such as cuprinol, or make a metal or plastic liner. Polyethylene sheeting stapled in place will last at least one season. If the box is much larger than one cubic foot use two-inch lumber and use rustproof nails to avoid stains. The exterior may be left natural, stained or treated with a beeswax finish.

Rough lumber tubs can be sanded or burned with a propane torch and the charred wood etched away with a stiff wire brush. However redwood and cedar left untouched gradually will weather and become more attractive over the years.

Exterior plywood can be used stained, painted or covered with cedar shingles, stick-on material—even carpeting—depending on what sort of decor you have in mind. A simple plant-

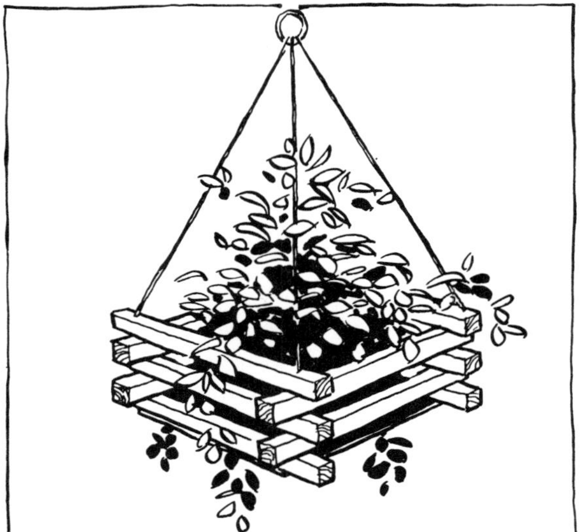

Hanging log-type planter.

er of exterior plywood is easy to make and can be capped and ribbed with strips of redwood. The exposed plywood can be stained, painted, or burned and wirebrushed for a pleasing effect. Lumber companies will cut the panels for you and then the assembling is relatively simple.

Wooden hanging baskets are easy to make, too. All they require is a base, usually a square of exterior plywood, and a number of "logs" or slats which are threaded on a wire coming up from each corner of the bottom piece. If made of redwood slats no treatment is necessary. Other woods should be treated with preservative or stained and waxed.

Making your own concrete planters is a more difficult task but is quite rewarding. An inner as well as outer form is used and the concrete made lightweight by the addition of perlite or vermiculite. Don't forget the drain holes.

If you have access to a potter's wheel and a kiln you might like to try your hand at making your own terra cotta pots and containers. Or if a potter is nearby you might have special containers made to order.

Other things adapt easily to plant containers. You can make two small tubs by sawing a nail keg in half and then drilling a few holes in each bottom end, or the keg can be used in its entirety as one large container. The inside should be treated with preservative or a polyethylene sheet liner stapled into place. A wine cask or whiskey barrel cut in two will make two nice large tubs. The heavy barrels used to contain liquids are preferred over more flimsy "potato" barrels.

Sturdy wooden boxes or crates can be adapted with the same precautions for preserving the wood and even a half-bushel basket will serve if lined first with polyethylene.

Metal containers such as galvanized pails, buckets, washtubs, sap buckets, waste baskets, large cooking pots (an old enameled hot water bath canning pot is excellent), and many other metal receptacles can be adapted—even an old tea-kettle. Drainage holes must be drilled or punched in the bottoms, and unless rustproof they should be coated inside with asphaltum (the black asphalt emulsion used to seal tree wounds). Spray-paint the outsides with rustproof lacquer or enamel.

Deserving special mention are the disposable aluminum loaf tins, bread pans, roasting pans, pie tins and TV dinner trays. The deeper of these make excellent quarters for cuttings or seedlings, the shallow ones useful drainage-catching saucers to place under other containers. I use a large aluminum turkey-size roasting pan to bottom-water my seeds flats and peat pots.

Milk cartons from half pints to gallon size can be cut down easily to form inexpensive, disposable containers for all sorts of uses, as can new thin but hard plastic jugs. Holes must be punched or cut in the bottoms, though, for drainage.

Today, many household containers are made of plastic and all can be adapted to plant container use. The list is almost endless and would include plastic pails, baskets, buckets, child's sand pails, ice cream containers, bread boxes, sandwich boxes, sweater boxes, dishpans, waste baskets, trash cans and plastic bags. Even plastic strainers and colanders make attractive hanging baskets.

CHAPTER EIGHT

What's in the Pot

In your container, besides roots of course, is the soil—something that too often is taken for granted. Much maligned, frequently called dirt and little understood, this part of your container-plant-growing system is just as vital to the healthy life of your plants as the air and sunshine.

The soil in each container provides the plant with a continuous supply of vital water, plant food and an anchorage for the roots to support the plant. Soil when in good tilth also provides air for the roots so they can "breathe," and as well with the proper temperature and humidity conditions conducive to good plant growth. And good soil is alive with millions of beneficial microorganisms that play an important part in furnishing nutrients to the plant.

Taken all together, what's in the pot actually is a micro-environment of mineral, vegetable and animal life that furnishes the plant with all the things it needs to supplement what is provided by the air and sunshine above.

I know that some horticulturists object to the term "dirt," but the terms dirt and soil are interchangeable. And I like the sound of "just a plain dirt gardener" because most who call themselves that in fact are most excellent gardeners, and really know that dirt is just not plain.

Dirt or soil is a very complex thing. It actually lives and breathes and even gets old. But it can be rejuvenated and brought back to a full, healthy life again and again. The reason, of course, is because one of soil's most important components is organic matter.

Because of its "living" characteristic soil can be managed or "cultured," to use a very old term. You can add things, take things away, manipulate its physical, chemical and biological activities—all for the good of the plants or, through neglect, to their detriment.

Let me describe briefly a good container soil before discussing its management. It has certain characteristics of weight, bulk or density, structure, color, temperature, moisture, acidity or alkalinity, nutrient content, and the relative availability of these plant foods to the roots, water content and air content. The weight of a soil is largely due to its mineral and water content. Organic matter and the plant roots add very little to the weight.

The proportions of mineral and organic matter, water and air, and how they are arranged or "structured" in the soil, determine its bulk and whether the soil is loose and friable, or "heavy," dense and compact. The color of a soil is derived in part from the various minerals and chemical substances, the amount of humus, water and air present.

Acidity or alkalinity frequently is worried about, and is little understood. This important soil characteristic is caused by the chemical nature of the minerals, organic matter and plant nutrients present, and the activities of the roots and other living components of the soil. The nutrients or "plant foods" in your soil are derived from the minerals, organic matter and the fertilizer you add.

The water content of a soil is determined by how much water is added and by how much drains out, versus how much is retained by the soil. How well a soil gets rid of excess water and yet holds on to what the roots need is related to the number, size and arrangement of the pores or open spaces and channels of the soil, which in turn are determined by the structure of the soil. The soil air (with oxygen of principal concern) also is determined as to amount, as is its movement into and out of the soil, by the soil's structure. Humidity refers to the amount of the water vapor in the air spaces inside the soil and to that of the air immediately above the soil.

The temperature of the soil usually is dependent on the temperature of the surroundings where the container is placed, but often is modified by the insulation qualities of the container, the water and organic matter content, and the color of the soil.

You may have noted that in describing the attributes and characteristics of container soil—or for that matter of any soil—I have used repeatedly the terms mineral, organic matter, soil structure.

The mineral content of the soil in your container depends on two things: what was present in the original soil mix, and what you added later when "feeding" or fertilizing the plant, less what is washed out by water moving through the soil, and what is removed via the plant roots. Soil minerals were derived originally from the mineral content of the rocks that went into the making of the soil thousands of years ago. These tiny rock and mineral fragments constitute the sands, flour-like silts and clays of natural soil, and when dissolved in the soil water through various physical, chemical and biological processes they furnish many of the plant nutrients.

The organic matter part of your potting or

container soil is made up of literally millions of microscopic organisms, both alive and dead, that include all classes of beneficial soil bacteria and fungi; the living roots of the plants themselves and plant by-products. These organisms contribute greatly to the vitality of the soil and help make many of the minerals available to the plant roots. They also are largely responsible for the development of humus. Humus is defined as the fully-decomposed residue of all sorts of organic matter, the end products of microbial activity. Humus is a very complex and little-understood substance, usually brown to black in color.

Humus and organic matter in various stages of decomposition contribute to the dark color of soil. More importantly organic matter serves as a source of energy for most of those beneficial microbes; has a great capacity for absorbing plant nutrients; helps sticky clay soils to become more friable; helps dry, sandy soils absorb water; aids in sticking together the tiny mineral fragments of the soil to create stable porosity, which facilitates the movement of air and water in the soil.

Outside of the fact that organic matter and humus assist beneficial soil organisms and help create conditions of good tilth in a soil, perhaps its most valuable contribution is its great capacity for making certain nutrients, such as calcium, potassium and magnesium, more available to plants. Organic matter itself is a major source of such plant nutrients as phosphorous and sulfur. It is also thought that humus furnishes certain growth-promoting and vitamin-like substances that stimulate plant growth.

The third term, soil structure, refers to the arrangement of the individual mineral grains into clumps or aggregates. These groupings or granules of soil minerals are what make up those visible crumbs of a soil in good tilth. Organic matter plays a part in their formation, the humus furnishing the "waterproof" glue, along with some sticky clay, to cement them together. Too much sand or clay and insufficient organic matter gives you a soil of poor structure. Organic matter functions as a "granulator" of the sands and clays, making soils loose, friable and productive. This looseness is important to good root development, water movement and nutrient availability. Thus a soil of good structure—in good tilth—not only can hold more water for the plants but is capable of allowing an unwanted excess to drain readily away. This in turn allows necessary air to move into the soil. Such a soil will have almost equal volumes of mineral particles and of the spaces between them.

Plant Food

Plant foods, or more scientifically plant nutrients, are as essential to plant growth as are air and sunshine. Other than those supplied mostly by air, these nutrients are furnished by the minerals in the soil, by organic matter and by what you add in the way of fertilizer and other soil amendments. These nutrients or essential elements in the soil must be present in forms available to the roots and usable by the plants.

Most container plants obtain the essential elements of carbon, hydrogen and oxygen from air and water. From the soil they derive nitrogen, phosphorous, potassium, calcium, magnesium and sulfur. Also they obtain, but only in relatively small amounts, the essential elements of iron, manganese, boron, molybdenum, copper, zinc, chlorine, and cobalt.

Other minor elements such as fluorine, iodine, silicon, sodium, strontium and barium sometimes improve certain plant growth. Very little is known about these elements, but to be on the safe side many organic gardeners use small amounts of kelp or seaweed to provide all of these minor but possibly essential elements.

The bulk of these elements is provided to the plant in the form of mineral or fertilizer additions to the soil, or in the form of organic amendments such as humus, composts or manures.

The plant foods or fertilizers you use commonly have the percentage content of their mineral nutrients identified on the label. They always list these nutrients in this order: percentages of nitrogen, phosphorous, potash, sometimes plus magnesium. For example a 5-10-10-2 nutrient analysis on the label would indicate that the fertilizer in the bag contained (in most states it's guaranteed) 5 percent total nitrogen, 10 percent available phosphate, 10 percent water-soluble potash, and 2 percent magnesium. Some companies list the ingredients from which the nutrients are derived, and these sources can be organic or inorganic, manufactured chemical ingredients or those derived from natural rock minerals.

The chemical fertilizers most useful to container-grown plants would include a number of analyses such as 5-10-5, 8-16-16, 10-10-10, 15-10-10, 16-32-16, 20-20-20. The last two are starter solution fertilizers. Fertilizers with all three nutrients present are known as complete plant foods. Fertilizers containing no nitrogen, such as 0-10-10, are useful in potting soil mixes. Others just with phosphate present such as 0-20-0, are for seed-starting mixes. Completely water-soluble formulas are best for container use. Commercial types available are mentioned on page 19. Certain fertilizers are formulated especially for acid-loving plants such as azaleas or rhododendrons, and there are others called "rose food," "African violet plant food," and so on.

Some fertilizers are called "organic" although they are truly so only if all their constituents are derived from plant or animal products residues, composts or manures. Some of the "organic" constituents, however, may be derived from natural mineral or rock deposits.

The true organic fertilizer materials include the following with their approximate nitrogen-phosphate-potash percentages in brackets [the symbol "t" indicates a trace may be present]: alfalfa leaf meal (3-½-0), cottonseed meal (6-2½ -2), dried blood (10-1½-½), fish meal (10-4-0), fish emulsion (5-2-2), soybean meal (7-1½-2½), bone meal (2-26-0), kelp (1-½-8) plus many other elements, wood ashes (0-1½-5), dried garbage tankage (6-8-1), sphagnum peat moss (1-t-t), activated dry sewage sludge (4-5-1).

Natural minerals, many traditionally recommended for specific uses, analyze approximately as follows: rock phosphate (0-25-0), granite dust (0-0-4), greensand marl (0-2-4).

Manures and composts usually are complete fertilizers in that the major plant nutrient elements are present. Though the percentages of these ingredients are fairly low, this should not discourage you from using well-rotted and specially-composted manures, for their other attributes are most important, especially for enhancing the tilth and water-holding capacity of your container soil. Examples of dried manures that you might use and their plant food percentages are: cow, (2-1-2), guano (9-9-2), poultry (5-3-2), sheep (½-t-½).

Some complete fertilizers and fresh manures are concentrated sources of plant nutrients, but use them carefully. Applied too close to the roots they can cause severe "burning." The use of water-soluble fertilizer mixes and starting solutions is preferable. Follow the directions on the fertilizer package or else fertilize at every third watering. Another method is to use one-third the normal amount and apply it to the containers as a very dilute solution at every watering.

Supply enough water, in either case, to each container so that a fair amount drains out the bottom. This will prevent fertilizer salt incrustation from building up on the surface of the soil. Whether fertilizing or not, some water should come out the drain hole each time to carry excess salts on through.

Fertilizer salts, iron sulfate and lime from hard water can cause floor stain. The containers could be moved elsewhere before watering or fertilizing, or suitable pans or saucers used under the pots or planters.

For acid-loving plants, such as azalea, blueberry, rhododendron, use a fertilizer mix containing an ammonium form of nitrogen such as ammonium sulfate. The special fertilizers for such plants should be applied according to the directions on the package.

If fertilizers are included in your homemade soil mix recipe preferably they should be of a slow-release commercial type, organic, or composted or dehydrated manures. These then can be supplemented later with water-soluble fertilizers.

Lime, a substance which provides a source of calcium (one of the essential plant nutrients), is more often suggested to correct acid soil conditions. A little bit of lime (see next section) can be added to seed-starting or potting soil mixes to ensure soil sweetness. Use two-thirds of a cupful of ground limestone per bushel of soil mix, or one-fourth teaspoon per four-inch pot, one teaspoon per seven to eight-inch pot. Lime isn't very soluble, so it should be mixed into the soil as much as possible. One application should be good for a year or so. If you use dolomitic limestone this will provide magnesium as well as calcium.

Shrubs, vines, ground covers and perennial flowers respond well to regular fertilization twice or three times during the growing season. Annuals and roses can use more frequent fertilization or a constant source of nutrients from slow-release materials. Apple, pear, flowering crabapple and other small ornamental trees do best with two to five cups of 10-10-10 or organic forms each year. For peaches, plums and cherries use half this amount. Citrus respond best to thrice-yearly applications of a five-percent nitrogen complete fertilizer plus a yearly application of chelated trace minerals.

Container-grown evergreens and conifers like 10-10-10 or similar complete fertilizers annually or every other year, at the rate of one to two cups per inch of trunk diameter.

One new way to apply fertilizer to container-grown plants is to use slow-release fertilizer tablets. These are placed down holes in the soil in the container. How many tablets for what plants and what sized containers depends on the nature of the preparation, so you should follow package directions explicitly.

Container Soil Mixes

If you have only a few containers to fill, probably it will be most convenient to purchase ready-prepared soil mixes for whatever your plants' needs call for. Such usually are sterile, with no harmful microorganisms or weed seeds present, and already they may contain the necessary fertilizer ingredients. A cubic-foot bag of prepared mix will be sufficient to fill six or eight ten to twelve-inch pots—more if the containers are smaller. Special mixes are available for acid-loving plants such as African violets.

It may be advantageous to mix your own, however, especially if large quantities are needed. A good soil mix contains equal parts by volume of loam topsoil, peatmoss and sand. You can substitute nitrogen-stabilized sawdust or ground bark or leaf mold for the peat, and substitute vermiculite or perlite for the sand. To be sure there are no weed seeds or objectionable microbes in the topsoil, you can sterilize it by baking it in your oven. (see page 15)

The above standard mix is excellent for most container uses, including starting seed and rooting of cuttings, although vermiculite alone will suffice for those operations. If you desire you can add a teaspoon of ground limestone per quart of mix to counteract any acidity from the peat, sawdust or leaf mold.

Some gardeners, and I too, add a teaspoon of superphosphate or bonemeal or a tablespoon of rock phosphate to the mix. This kind of fertilizer is all right to add to seed-starting mixes as well. Usually the fertilizer is added to the containers periodically after the plants are well established.

If you include other fertilizers in your potting soil mix, it is best that they be of slow-release, organic or composted types to facilitate even plant growth.

Potting and Re-potting

Getting your newly-grown transplant or rooted cutting, or new plant from its temporary nursery container into its permanent container home is not as difficult as some would lead you to believe.

The trick is to get the plants transferred with the least amount of damage to the roots, particularly not letting them dry out the least bit.

Select your new containers (or carefully washed old ones) for proper size. They should be slightly larger than those where the plants now reside. Cover the drain holes with pot shards (curved, broken pieces of old clay pots), flat stones, or small pieces of one-eighth or one-quarter-inch mesh hardware cloth screening. Put in a small amount of moist soil, then the plant or cutting, and fill in around its root mass with more soil mix. Press the soil down gently but firmly with the fingers to eliminate any air pockets. Fill the container up almost to the rim; then water. When potting into very large containers, tubs and planters, fill about halfway with the soil then soak with water, let the soil settle, then repeat with more soil and more water until the container is filled to within one inch of the rim.

If you're handling nursery plants grown in cans or plastic containers, slit the sides of the cans and spread open before extracting the plant.

In rejuvenating a rootbound plant it should be re-potted into a slightly larger container after first lightly scoring the root ball with a knife, as suggested below for nursery container-bound plants. If it is to go back into its original pot, tub or planter you can slice off a thin layer of the outer roots. Bonsai plants usually are root-pruned and replanted in the same container.

Sometimes it's difficult to remove a rootbound plant from its old container. If it's small turn the pot upside down and, with one hand supporting the soil, rap the rim against the edge of a table. For larger containers, water

Slit the nursery can.

Re-pot in larger container.

thoroughly the day before, then lay them on their side and rap the rim sharply with a wooden or rubber mallet. Protect the container rims with a block of wood or cloth. Running a knife or long spatula around inside the container between pot and soil will help too. For large tubs and planters, try using an old hand saw the same way. Forcing water through the drain hole with your garden hose also may help loosen the root ball. After the root and soil mass has been loosened, carefully slide it out of the container. When lifting the root ball, support its weight with one hand.

Do all potting and re-potting in the shade, since direct sunlight will dry out the roots quickly. The fine rootlets of seedlings and rooted cuttings are particularly susceptible to drying, and many have been lost through failure to keep them moist.

If the root mass appears potbound (rootbound)—that is, a mass of roots are exposed on the surface of the ball and the whole mass has assumed the shape of the container—score the rootbound ball with a knife. Make at least four vertical cuts from top to bottom with a knife, cutting into the mass about one-fourth-inch deep.

Position the new plant in its new container so that the soil line on the stem is now about an inch and a half below the rim. Fill in and around with soil. Seedling transplants should be planted not much deeper than they grew in the seedbed (see Chapter II). Some transplants and rooted cuttings usually won't have much of a root ball, and if so you can fill the new containers loosely with soil, poke in a suitable size hole with a dibble or your finger, into which the transplant or cutting is inserted carefully and gently. Then press down gently with the fingers, firming the soil mass and adding more potting soil mix if necessary to fill the pot.

In re-potting, to shift a plant into a more decorative container or if the plant has become rootbound, you should select a slightly larger container if possible. However, the same sized pot or planter may do, and in some instances you can re-pot right back into the same container.

But when a plants' roots have outgrown their container—the symptoms are reduced top-

Rap the edge of the tub to free the plant.

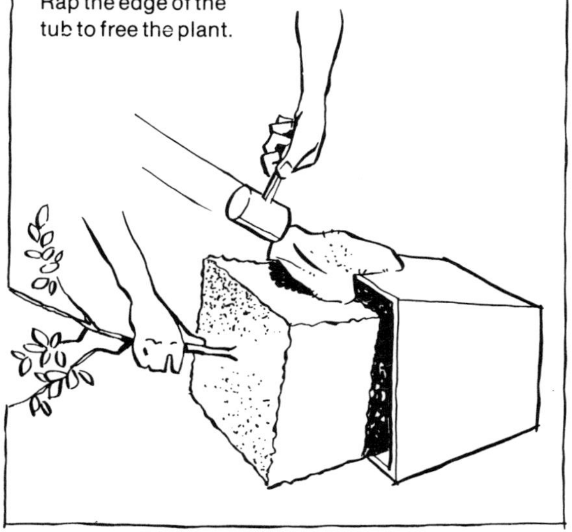

Scoring a root-bound plant.

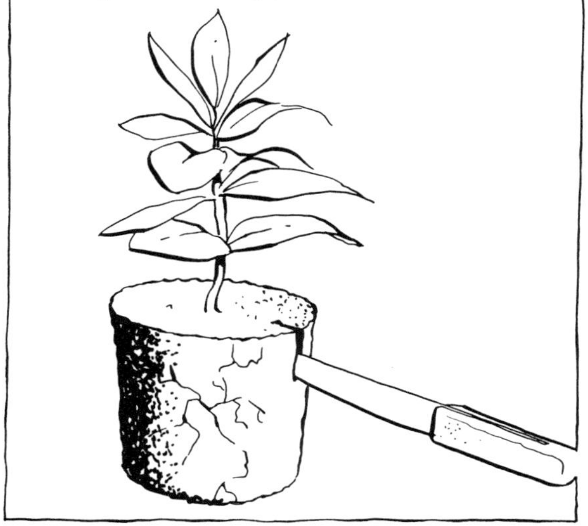

growth or vigor and roots poking down through the drain holes—it's time they were given a new home, or at least rejuvenated with some root pruning and new soil. Some exceptions to this rule, where crowded roots benefit blooming, occur with Kafir lily, agapanthus, some succulents, often roses and some annual flowers.

Pruning and Shaping

Pruning container-grown plants involves more than the root pruning noted above under Potting and Re-potting. It includes modifying the shape of the above-ground parts of the plant as well.

Pruning back top growth is done primarily to modify the size of the plant, to keep it in harmony with the size and shape of its container. Pruning also is done to enhance flower and fruit development, to change the shape of a plant into a different form—such as for standards, topiaries, miniaturizations and the exotic forms of bonsai.

With succulent annuals and perennials all that is needed to limit top growth and to encourage more side branches and a proliferation of flower and fruit is to pinch out the growing tips with your thumb and finger. If the growth is getting woody use a sharp pruner, and make the cut down next to a leaf or lateral bud. Leaving stubs only invites disease problems. When the terminal bud (or buds in the case of multiple-stemmed plants) are cut back, growth stops at that point, and lateral or side branch buds will develop. When many multiple stems are present, cut some back to the ground line. A bushy plant on one main stem looks more pleasing than a scraggly one growing many slender and over-tall stalks.

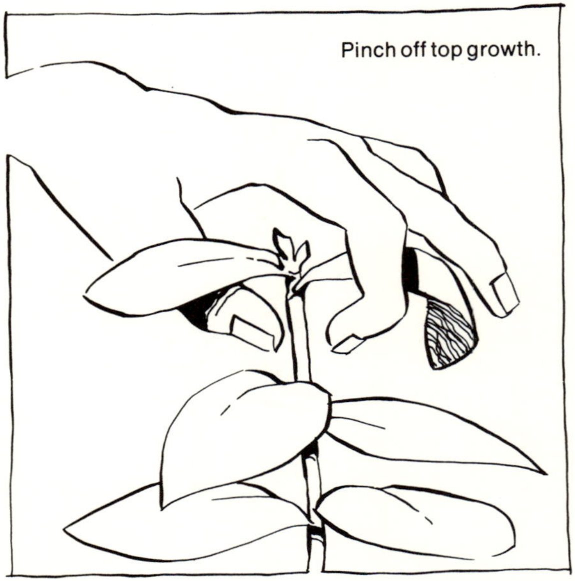

Pinch off top growth.

Pruning dwarf citrus usually is not necessary except for the periodic removal of dead twigs and branches, and perhaps a live branch if it crosses over another. Other dwarf deciduous fruits, however, need severe pruning, thinning out and cutting back of the top branches (heading) to maintain desirable size and form. Apple, apricot, cherry, and pear dwarfs should be maintained at about 30 inches in height, although some dwarf cherries can be cut lower. Peach and plum can be maintained as short as ten to twelve inches tall. It is easier to pinch off the succulent buds and shoots during the growing season than to wait to prune woody growth during winter.

Shaping container plants for form and character takes a bit of judgement on your part. In general, you prune so that the shape and appearance of the plant is in harmony with its container and is pleasing to you. Exactly where

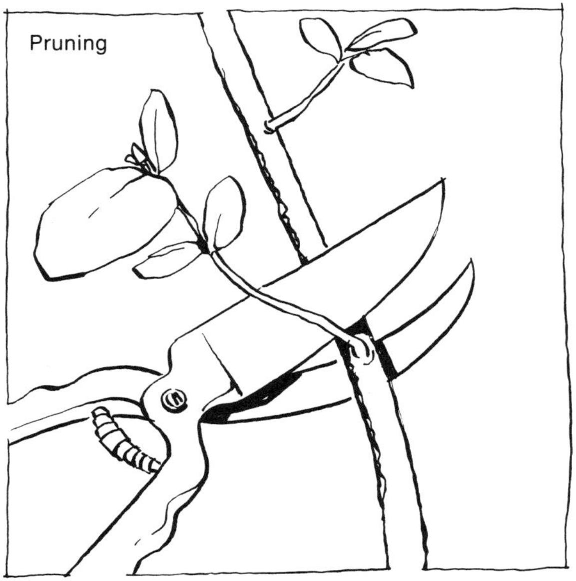

Pruning

Pruning dormant plants.

the cuts are made and how far back into old wood depends greatly on the kind of plant you're pruning. Some suggestions for careful pruning are: Don't just shear leaves, cut back branches; eliminate straggling, twiggy growth; cut out a branch that crosses another; cut close to the next bud down the stem to avoid unsightly stubs; use sharp pruning shears and make your cuts clearly, at right angles straight across the branch and not at a slant. Dormant season pruning encourages new branch growth; summer pruning encourages fruit spur development.

Pest Control

Sometimes insect pests will get into your containers, and pillbugs and slugs may be hiding under the pots. Propping your containers up on bricks or blocks of wood will eliminate much of this problem. In cleaning out used pots, be sure no slugs are hiding in the drain holes. One way to keep slugs from entering the pot via the drain hole is to cover it with a piece of window screen.

Usually, since you used sterile soil or potting soil mix to begin with, you won't have many soil pests to contend with. If you suspect their presence, however, try dunking the containers under water for an hour or so. You may be surprised at the number of soil insects that will float up. They then can be washed away easily, and insects or no, this is a good alternate way to water your plants.

When insect pests menace the leaves of your plants, the first thing to do is gently or forcibly—it depends on the nature of the plant—spray or syringe them off with water. Delicate leaves can be syringed, some wiped with a wet cloth. This method works especially well to get rid of spider mites and aphids. Only as a last resort use insecticides.

When employing insecticides on container plants be sure that you're using only indoor sprays or dusts on indoor plants, and outdoor pesticides on your porch or patio plants. This information is found on the label. If you use any pesticide on vegetables, herbs, or fruits, make absolutely sure that the kind of plant you're treating is named on the label on the package. Be sure you mix and apply according to the manufacturer's directions all pesticides, which now require approval by the U.S. Environmental Protection Agency. Store them, preferably locked up, out of reach of pets and children.

Weeds often are called pests, too, but usually they aren't much of a problem when growing plants in containers. Sterile soil and potting mixes have eliminated most weed seeds, and the composting of manures tends to do this, too. Any weeds that do sprout up in your containers can be pulled out easily by hand, and the sooner you do this the better. Pulling large weeds often damage the roots of your plant. The smaller the weed the easier it is to pull, and watering the container well before pulling the weeds will help loosen the soil and make the job easier.

For years gardeners have used mulches to prevent the growth of weeds in their vegetables, fruits and flowers. A mulch of fine bark chips, unbaked terra cotta granules, even small pebbles, will be useful to keep weeds out of your containers, and will be attractive as well. Discs of fiberglass or a polystyrene foam felt-like material are commercially available. Slit them to the center and place around the plants' stems to cover the soil surfaces.

A mulch not only prevents weed growth by excluding sunlight from the soil surface but it also helps retain soil moisture, an important consideration for containers on a sunny patio.

CHAPTER NINE

Spring and Summer for the Handicapped

Container gardening can open new horizons for people who are prevented from gardening in the usual way due to infirmity or handicap. Those who are confined to the wheelchair or just can't walk, bend or stoop with ease now can grow their own plants for fun, beauty, food or flavor. Containerized plants can be arranged for accessibility and easy reach, while supplemental equipment is available to augment and extend one's gardening abilities.

In another aspect container-grown plants, because of their mobility, can be transported to those confined in bed, and with a suitable bed tray arrangement they actually can work with them to some extent.

Not only do foliage and flowering plants fulfill this need, but so do such miniature arrangements as dish gardens, bonsai or gardens-in-glass. A terrarium or a bottle garden is a never-ending source of enjoyment to a bedridden youngster. And for those confined to the house, container-grown plants and a fluorescent light setup means a chance otherwise denied to grow plants indoors any season of the year.

For Those Who Can Walk

One has a choice of gardening situations depending on his abilities or limitations. If he can walk around but has difficulty bending to reach the ground, minigardening with special long-handled tools and other no-stoop devices may be the answer. Plants grown in small, light-weight containers will be easy to handle. Better yet, containers arranged on a moderately high wall, table, bench or platform will eliminate the need to lift the pots to working height or for stooping.

Many tools are available to extend a person's reach or some can be adapted for such use. Flower-gathering tongs are available at some garden centers, as are long-handled scythettes, scuffle hoes, narrow rakes and long-handled dibbles. Of course many of the smaller common garden tools will be quite maneuverable and useful.

If one's problem is not being able to bend down all the way, a portable kneeling stool will aid in gardening. These are available at hospital supply houses and come with padded benches and easy-grip handles. One can be made with a few pieces of wood for handles fastened onto a sturdy apple box.

Still the best solution is to bring the garden up to working height, rather than to try to get down to plant level. Lightweight plastic or wooden containers are quite portable when a lightweight potting soil mix is used. Lots of perlite, please.

Another feature of container gardening that can be worked to advantage is to place pots or other containers on a wheeled cart so they can be perambulated from house to patio or elsewhere. Wheeled carts designed expressly for this function are available from some florist supply houses, or a hospital medicine cart will adapt to this use. A portable garden also can be made by attaching large casters or small wheels to the legs of a small wooden table.

For Those in a Wheelchair

For those confined to a wheelchair the containers can be raised beds of a suitable height, or

Portable kneeling bench.

wooden-rimmed tabletop gardens with a collection of smaller containers well within reach. The key is easy access by wheelchair to the "garden" and easy reach to the plants without undue stretching.

In the patio or garden, paths should be wide enough and their surface smooth and hard for easy chair manipulation. The beds, raised on cement block or brick, should be at a convenient level for easy access to the soil and plants from the wheelchair. The beds can be centrally located within the patio or pathed garden, around the perimeter or both, if the size of the available area allows. A very attractive garden can be laid out in this manner and can be maintained easily from the wheelchair.

Plant beds raised to wheelchair working height with stone or other masonry, are more or less permanent and often require working sideways from the chair. However, container platforms built of wood with legs of a proper height will allow the front of the chair to fit underneath so that one can work from a frontal position. Such a raised platform should be sturdy enough to carry the weight of pots and soil. Two-by-six lumber is excellent. Decay-resistant woods such as cypress or redwood topped by six inches of soil usually is sufficient.

Beds containing soil should have cracks between the planks or holes bored in the bottom to allow excess water to drain away. When working with plants from a frontal position with the chair partly underneath the garden, the seated gardener should protect his lap with a sheet of plastic.

If plants are grown in containers, they should be placed also on the raised bed for easy wheelchair reach. And containers instead of soil in a bed makes gardening that much easier.

Miniature or midget varieties of plants are especially useful and adaptable to wheelchair gardening. Plants which will grow out of reach should be avoided. Selecting plants easy to grow, those which require a minimum of manipulation and re-potting, is most practical. Terrariums, midget gardens and bonsai also are excellently suited to wheelchair gardening as is the construction of bottle gardens and berrybowls.

For Those in Bed

Those who are confined to bed, depending on their capability, can enjoy container-grown plants in different ways. If propped up, one can enjoy working with small container-grown plants and do other things, such as plant seeds and transplant seedlings. If the light garden is nearby, its management can be overseen and its visual benefits enjoyed. Constructing berry bowls or bottle gardens may be well within the capabilities of the bedridden.

For the person unable to do much more than observe from bed, a nearby light garden or a small terrarium or garden-in-a-bottle on the bedside table can be a source of great pleasure. (see Chapter XII). A simple bottle garden made with a few ferns and moss and a partridge berry or two brings real enjoyment to a child in bed.

For the Exceptional

A very useful aspect of container gardening is the ease with which plants can be grown. So the child or adult handicapped other than physically can gain a great deal of experience and enjoyment through this form of gardening. Depending on their capabilities—and often this can be developed—they can grow a great variety of plants in containers.

If they start out with easy things, such as planting seeds, and progress through the various stages of growing plants under lights, I believe you will be pleasantly surprised at what they can learn to grow, and the rewards they derive from the experience of achievement.

CHAPTER TEN

Vacation and Summer Camp Gardening

Growing plants in containers and gardening under lights can be a great boon to you in the summer, since you also can take your plants with you when you go on vacation. Plants in containers are adaptable to moving great distances. You can pre-plan your outdoor vacation home garden, start the seedlings under lights ahead of time at home, transport the young transplants to your spot at the shore or mountains, and when you arrive have an instant garden all ready to grow.

These features are quite adaptable to the summer camp counselor's use as well as for teaching plant life, biology and ecology. All are made easier and more meaningful by having growing plants readily at hand to show and work with.

Vacation and Your Plants Stay Home

Many people feel they can't grow plants in their homes for fear they will be neglected while away on vacation, but plants in containers make it all much easier.

If you don't wish to take your plants along (or maybe you have too many to fit in the car), you can leave them where they are under automatically-controlled lighting and watering; put them out on the porch or patio where natural light will do (and a neighbor can tend to their watering); or you can sink them temporarily, pot and all, into an appropriate spot in your garden and trust to nature's rain and sunshine for their care.

Automatic Watering

One way to provide for watering container-grown plants is to use one of the commercially-available devices or containers (some sources listed in the Appendix) especially designed for this function. But if you have a great number of plants this could be quite an expense.

For a short time a few containers can be immersed in a pan or in deep saucers of water or even in the sink or bathtub. Since the water will evaporate about as fast, perhaps faster, than the plants use it, this method is useful at best only for periods of short duration. However, with a little trial and error you can adjust the faucet to drip just fast enough to maintain a suitable water level. Make sure your sink has an overflow drain, just in case.

Descending wick watering.

Another method, the descending wick, is useful for a single container or a group of them in a tray or pan. Suspend a pail of water overhead and allow a short length of soft rope—braided cloth works well—to run from inside the bottom of the pail, over the rim and down to the container. Water will be fed by capillarity and drip down to the soil in the pot, keeping it moist. This method's shortcoming is that it hasn't capacity to keep a large number of containers watered, unless you make several such setups. And it works only as long as the water in the pail holds out. A test run will tell you how long your pail will last.

Another wick-watering method involves suspending the pot over a pan of water. One end of a cylindrical wick of asbestos, fiberglass, braided soft cloth or soft rope is tucked firmly into the bottom drain hole of the pot, and the other end is placed in the water. The best way to do this is to knock the plant out of the pot carefully, tie a knot in the wick and thread the free end down through the drain hole leaving the knotted end inside the pot. Replace the plant and suspend the pot over the water can with the wick dangling down into the water.

Automatic watering devices can be constructed with time-clock controlled electric-solenoid water valves, adjusted according to how often and how much water is required. Such an automatic system requires the assistance of a licensed plumber and electrician in some localities. The initial expense would be moderate, but the long-run benefits great. Of course you would be able to utilize its advantages not only during your absences but all year long.

One system that is only good for a few days and at best for a week or so, is just to water your plants well, giving each container a good soaking. Then slip a plastic bag of adequate size down over the entire plant and the container. Tuck the open ends of the bag under the container or tie closely around the pot with a string or rubber band. Moisture given off by the leaves by transpiration will stay within the bag, condense and drip back onto the soil.

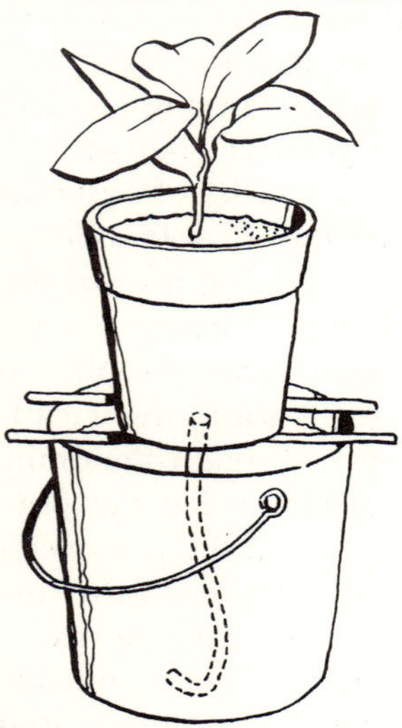

Suspended wick watering.

Vacation and Your Plants Go With You

Of course you may wish to take your container-grown plants and perhaps a modest-sized light setup right along with you when you vacation. If there aren't too many and you have space available in your vehicle, this is fine. It eliminates any need for plant sitters at home.

To transport a number of plants in containers, line a suitably-sized sturdy carton with polyethylene sheet plastic—drycleaner bags will do—and pack the pots inside this in a bed of damp sphagnum moss. This way it is easy to handle a number at a time, and it's easy on the plants during the trip. This method works well also for transporting seedlings or young plants in peat pots, and it applies as well to a number of annual vegetables and flowers not usually grown as transplants. Even corn, melons and squash can be started in peat pots to be set out upon your arrival.

When planning an outdoor vacation garden

select the varieties and kinds of plants that are adapted to your vacation climate and those that will mature within the period you will be there. The local county agricultural extension agent near your vacation site will be able to advise on growing seasons and locally adapted plant varieties. When vacation is over, reverse things and bring them back home with you. The plants already in containers simply go back into the carrying box. Others you may dig up and pot to transport. See page 23 for details.

Summer Camp and Classroom Gardening

The portability of container-grown plants makes it possible to have live plant materials of the proper size and age for use both in summer camp and winter classroom educational programs.

In Summer Camps

Nature counselors and others of the camp staff concerned with plants can adapt many of the suggestions noted earlier in this book. Flowers, fruits, vegetables and other plants particularly adapted to demonstrating certain biological and ecological concepts, herbs and even a whole vegetable garden for the camp kitchen can be pre-planned and started ahead of time.

Some of the many demonstrations and experiments applicable in summer camp include showing which seeds germinate best in dark or light, how seeds sprout, how plants bend toward light, how green chlorophyll disappears under taped-on lightproof patches, how plants grow and function, which parts of plants are edible, how flowers work and how plants react to pollutants (such as bean plants exposed to automobile exhaust.)

All these programs develop in youthful campers an awareness and understanding of plants, how they grow and the many ways that plants are useful.

Winter Classroom Gardening

The classroom teacher can move container plants into the classroom for a single day's demonstration or grow them from seed to flower in the classroom for long-term projects. They can be moved from room to room and be taken home for maintenance over the weekend or during longer vacation periods.

The wide selection of seeds, seedlings and plants available to the teacher not only represents examples of useful kinds of flowers, fruits, vegetables or foliage plants but serves for other botanical or horticultural experiments.

Mobility of Classroom Materials

Numbers of small containers may be grouped in trays or flats and be moved easily by hand or on carts. Larger plants, perhaps a fruit tree in flower to illustrate pollination, can be moved on low wheeled carts. Plants on a windowsill garden or under a light setup can be carried to the student's desk for close-up observation and study, then back to their permanent garden spot.

One should not overlook the double value of adapting commonplace receptacles as plant containers, in the ecological use of something otherwise discarded.

Plants for Classroom Use

The kinds and varieties of plants used or grown in the classroom are determined by what uses are to be made of them. Those that grow most rapidly and are easier to manage, and those that prosper on windowsill situations, are better to use, since school time is limited.

If the teacher wishes to demonstrate seed vitality, dormancy, the need for scarification or light for germination, embryonic development and seedling emergence, this would call for seeds and seedlings. For an exercise on chlorophyll development or flower structures mature plants could be called for. The decision on what to study also determines to a large extent the sorts of containers and light setups needed.

The project at hand also will determine what type, kind or species of plants are needed. Botanical studies might well call for examples of different plant families, genera or species. Physiological considerations might require plants with differing capacities, functions or systems. An obvious choice of plants would occur to demonstrate which plant parts are edible, and the different leaf, flower or fruit structures.

It is not within the purview of this book to describe the many experiments facilitated by the use of container-grown plants. Suggestions noted on (page 94) for the summer camp counselor adapt well to classroom use. The botanical study of plant parts, structure and function, photosynthesis, nutrient uptake, needs and deficiencies, or simply how a seed comes up—all are facilitated by container set-ups.

CHAPTER TWELVE

Summer in a Bottle

A special kind of container-gardening finds the whole plant, not just the roots, residing in the container. Terrariums filled with plants, berry bowls or simple arrangements of mosses and ferns tucked inside glass jars fit this category.

There are two kinds of containers, or perhaps I should call them container systems. One has the top open and exposed to the outside atmosphere; the other is closed more or less tightly. Open-top terrariums or uncovered round berry bowls are examples of the former, while an arrangement of dieffenbachia, pipsissewa, ferns and moss in a tightly corked clear glass decanter is typical of the closed systems.

The covered garden-in-glass is self-contained, somewhat like the "closed system" of a space capsule where the air and water are continually refreshed and recycled. The plants will rejuvenate the air continually in the covered jar or bottle through the process of photosynthesis and respiration. Moisture, given off by the leaves in the transpiration process will condense on the inside of the glass and trickle down again to the roots. With the proper balance between soil, plants and moisture your container will need to be opened almost never again.

Three types of gardens can be made in either open-topped or covered jars and bottles: woodland, desert, or tropical—depending on the kinds of plants used. Both tropical and woodland plants do well in either type of container system but desert plants do better in uncovered containers. Tropical plants are best in covered ones.

Containers

Rectangular glass terrariums, fish tanks, plastic terrariums, fishbowls, candy jars, cookie jars, Mason jars, mayonnaise, pickle or peanut butter jars, brandy snifters, beer steins, glass cookware, gallon and half-gallon jugs, wine bottles, decanters, medicine and pill bottles, laboratory glassware, baby bottles, milk bottles, miniature bottles—almost anything glass or plastic that you can see through and can get plants into are adaptable.

Those that cannot be covered easily, those with large open mouths or tops perhaps are better suited to use for open-top containers. Those that have covers, tops, lids, stoppers or can be fitted with corks, work best for sealed, self-contained systems.

Once you get into bottle gardening and discover what fine gifts they make, a great deal of your pleasure will come from searching out odd and unusual glass containers. Interesting bottles and jars, particularly if antique, add to the value and enjoyment of the gift.

If you are working with children or trying this gardening technique for the first time, I would suggest you start with containers having rather large mouths or tops and which are not too deep.

Pint canning jars, mayonnaise and pickle jars, and instant coffee jars are easy to work with, and if your creations are not up to expectation they are easy to empty and start anew. Many of the children I have helped build gardens in a bottle quickly become adept and are eager to try their hands with challenging shapes.

Plants for Gardens-in-Glass

Plants suggested for tropical and desert terrariums are listed in Chapter VI. Some supplementing the woodland terrarium list are given below.

Mosses, particularly the sheetmosses, fern and sphagnum, are excellent for liners; other mosses for fillers. Use clubmosses or ground pine *(Lycopodiums)* and conifer seedlings for height accents. Ferns of all sorts, lichens (particularly Reindeer moss and red-capped British Soldier), teaberry, wintergreen or checkerberry *(Gaultheria),* the partridgeberry *(Mitchella repens),* ground ivy, mayflower *(Hepatica),* wood sorrel, violets, Springbeauty, wild thyme, Jack-in-the-Pulpit, Pipsissewa *(Chimaphila),* creeping cinquefoil, as well as the seedling conifers, are among the many wild plants for your garden in glass. When collecting mosses and wild plants, always leave some growing. Note that some ferns, clubmosses and other plants are "Endangered Species" and should not be collected.

Some of the small garden or house plants are good under glass also. They include Kenilworth ivy, strawberry begonia, lily-of-the-valley, periwinkle (or creeping myrtle), crocus, as well as the Snake plant *(Sanseviera),* miniature ivy, dumb cane *(Dieffenbachia),* small specimens of spider plants *(Chlorophytum),* Star-of-Bethlehem, and tiny forms of *Lobelia, Crassula* and *Lygodium.* Many of these are available at garden and greenhouse supply centers. Buy plants that are small but vigorous. Some of the wild plants also can be purchased from specialty garden supply houses, sources noted in the Appendix.

Materials Needed

1. Woods loam, leafmold or humus (the best is from the spot where the plants grew), coarse sand or fine gravel for drainage, and charcoal bits to keep the soil sweet, if you don't use a store-bought planting soil mix containing activated charcoal. 2. Gravel, particularly white aquarium gravel, or small stones or pebbles that will fit into the container for accent. 3. Miniature ceramic figures (if they'll complement your arrangement). I have found some excellently-proportioned plastic deer and other small animal figures in the Christmas decoration sections of department stores.

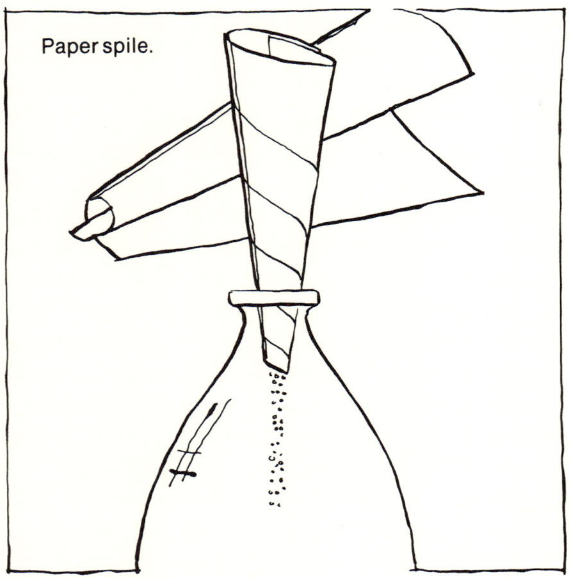
Paper spile.

In using figures or miniature structures such as bridges, you must use your judgment as to whether they really complement an arrangement or not. Sometimes the figures only distract or are unharmonious and the garden would be better off without them. On the other hand, a tiny fawn lying next to a miniature standing doe in a setting of ferns and mosses may add immeasurably. As one of the children said, "they really make it come alive."

Tools and Equipment

First of all, the tools needed for collecting and preparing your plant materials: A small trowel, plastic bags for carrying and a pair of old scissors for trimming mosses and ferns.

Next, the equipment for getting things into the jar or bottle: A small funnel, a rolled up or spiled paper tube or a short length of rubber or plastic tube—even a peashooter—to get the sand, soil, gravel and charcoal bits, into the container exactly where you want them. The tubes can be used by themselves but work better as extensions to the funnel. I made a very good tube by cutting a foot-long section from the hollow plastic handle of a child's toy mop.

For inserting plants, bits of moss and other objects a pair of quite long and narrow tongs is useful—surgical forceps, photography tongs or even lobster tongs. Another useful tool that holds on to things is a flexible mechanical finger-like instrument (see Appendix) with a plunger on the handle that activates two gripper jaws at the other end of a long flexible shaft.

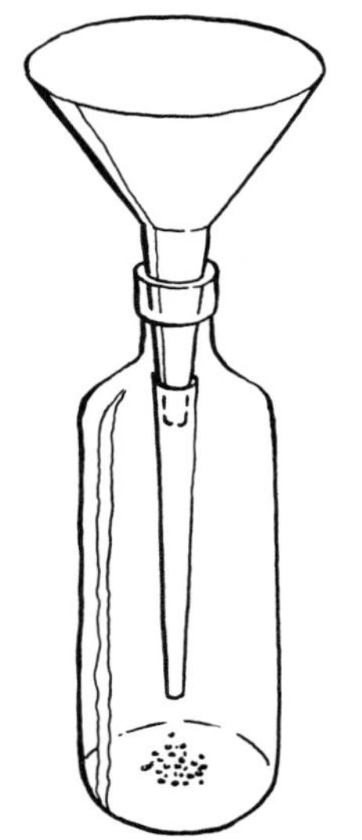

Extended funnel.

The other technique for inserting plants is to spear them on the end of a long narrow instrument such as an olive or fondue fork, icepick, chopstick (a pair of chopsticks also can be used to hold objects). Short forks can be taped to a dowel or pencil to obtain sufficient lengths to reach to the bottom of tall bottles. For short bottles a sharpened pencil works fairly well.

I find that a 3/16-inch dowel with one end kept sharply pointed in a pencil sharpener works best of all. The dowel can be reversed and the squared-off end used for tamping things into place.

To remove objects impaled on the pointed dowel I made from coathanger wire a long straight probe with a small C-shaped hook at right angles at one end. The size of the hook is slightly larger than the diameter of the dowel. When the dowel is inserted into the bottle with something impaled on the pointed end and moved to the spot where I want it in the arrangement, the hooked end of the long wire then is slid down the dowel to push the plant off. Actually the hooked end holds the plant pressed into place while the dowel is pulled out. This combination of pointed dowel and wire hook also can be used to pick up and move plants about or to extract them from the bottle.

An atomizer-type sprayer or a small rubber syringe are excellent for watering. Spraying in the water gives you an opportunity to wash bits of moss and soil off the leaves of the plants that are inside the container.

For cleaning the interior of the container (and it's almost impossible to keep some flecks of moss from getting on the inside of the glass), pipecleaners, a small percolator brush or cotton swabs do well. You may have to tape them on extension handles of dowels or wire to reach deep bottles and out-of-the-way corners.

Planting Hints

Select plants with similar environmental requirements, particularly as to light and temperature needs. Insert a moss liner, green side down and facing out, piece by piece—in one

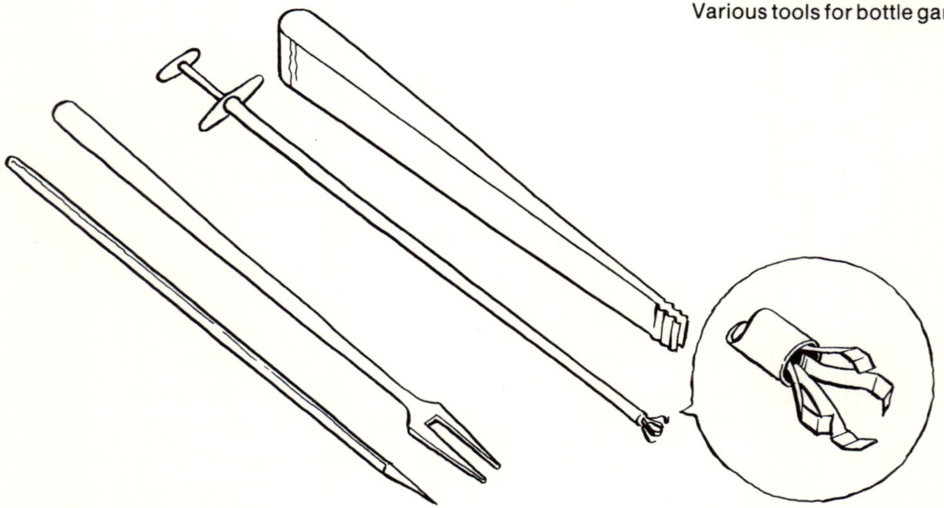

Various tools for bottle gardening.

sheet into larger-mouthed jars. For round-bottom flasks and hanging bottles the moss should cover the bottom completely for appearance, to conceal the soil, charcoal and roots from view. However, if the container is destined to sit where its underneath won't show, all you need line with moss are those parts of the sides where the soil otherwise would be visible. Then add gravel for drainage and a few bits of charcoal or dry soil-charcoal mix via the funnel and hollow tube. Your container now should be one quarter full (an eight-inch tall bottle would require two inches of moss, gravel and soil base.) Insert the plants, roots first, trimming them to size first if necessary. A small ball of moist soil or mud will help get the roots in place. Poke roots carefully into the soil, covering roots with soil and bits of moss. Arrange for simplicity, harmony and balance with the height to the rear or center. Don't overcrowd! Add bits of moss or gravel between the taller plants to help cover roots. Take your time, since haste really can waste a good bottle and fine planting materials. Next clean the inside of your glass container, brushing off sand, moss-flecks and mud stains; then water lightly and cover, cap or cork. The cover must fit tightly but is not permanently sealed.

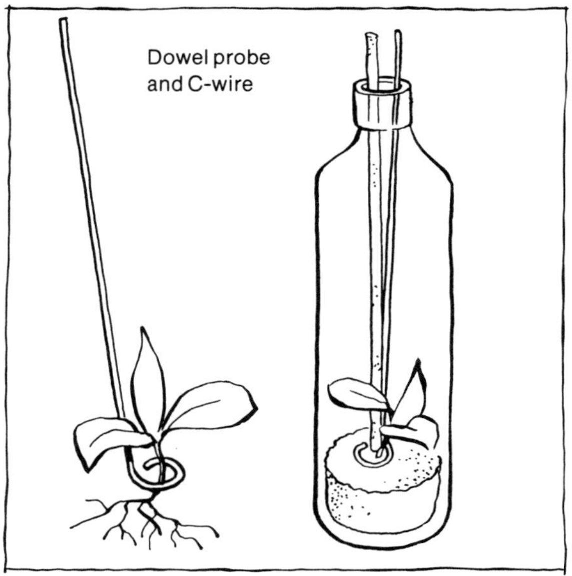

With a balance between moisture given off by the leaves and taken up by the roots, observable by just a slight misting of the glass, there is no need to open it up. Too much or insufficient water, the need to prune out excess growth or to replace a plant would be about the only reasons to open the top.

Put your bottle garden in a cool, moderately-lighted place for a few-day "setting in" period. Then it's safe to move it to stronger light in your room or wherever you wish.

Some Dos and Don'ts

Sometimes the balance of things isn't perfect, and you may need to make adjustments: Too much moisture condensing on the inside or decaying foliage? Open it up for a while to air it out. Yellow leaves? The roots are too wet and it needs less water or more gravel. Musty smelling? You forgot the charcoal! Shrivelling brittle foliage or brown crisp moss? Water, please. Bugs, gray molds or green algae climbing the glass? Interesting to watch but a tiny bit of household disinfectant spray might help, though maybe you'll have to disassemble it and start all over.

Don't overcrowd. Don't use fertilizer which will stimulate growth and cause the tiny plants soon to outgrow their quarters. Don't use grass in a bottle for the same reason. Don't over-water. Don't expose shady woodland plants to direct heat.

Do be sure your bottle or jar is shiny-clean, inside especially, before you start. Don't mix desert, tropical or woodland plants together in the same containers.

Gardens in glass work out well with a light garden setup, particularly where a single lamp unit is illuminating a specimen plant in a container, or the gardens can be lighted on their own in cases where the plants used respond well to such conditions. Special lighting is not a requirement, however, where low-light intensity woodland ferns, mosses and other plants are assembled.

Such gardens in a bottle need minimum attention and make excellent gifts for those confined to the sickbed. Children in particular gain great delight just watching tiny ferns and mosses living inside a bottle—especially when a tiny insect hatches out and flies around safe inside.

Appendix

Some Sources of Plants, Supplies and Equipment

Abbot's Nursery, Rte. 4, Box 482, Mobile, AL 36609—*camellias*

Alberts & Merkel Bros, Inc., P.O. Box 537, Boynton Beach, FL 33435—*orchids, bromeliads, tropical plants*

Leyse Aluminum Co., 203 Ellis Street, Kewaunee, WI—*aluminum planters*

Archias' Seed Store, P.O. Box 109, 106 E. Main Street, Sedalia, MO 65301—*vegetables, fruit, herbs*

Brighton Byproducts Co., Inc., P.O. Box 23, New Brighton, PA 15066—*tools, supplies, auto-waterers, containers*

Brookstone Co., 13 Brookstone Bldg, Peterborough, NH 03458—*tools, supplies*

Buell's Greenhouses, Eastford, CT 06242—*gloxinias, african violets*

Burgess Seed & Plant Co., Inc., 67 E. Battle Creek Street, Galesburg, MI 49953—*plants, seeds & supplies*

Burnett Brothers, Inc., 92 Chambers St., New York, NY 10007—*vegetables, herbs, and supplies*

W. Atlee Burpee Co., Philadelphia, PA 19132, or Riverside, CA 92502, or Clinton, IA 52732—*vegetables, herbs, dwarf fruit, supplies*

Burrell Seed Growers Co., Rocky Ford, CO 81067—*vegetables, peat pots*

Cactus Pete's, 5454 Valley Blvd., Los Angeles, CA 90032—*cacti, succulents, epiphyllums*

Concord Woodworking Co., W. Concord, MA 01781—*redwood planters*

Corning Glass Works, Corning, NY 14830—*glass containers*

Craft-House Mfg. Co., Wilson, NY 14172—*lighted stands*

DeGiorgi Co., Inc., Seed Growers & Importers, P.O. Box 413, Council Bluffs, IA 51501—*vegetables*

P. De Jager & Sons, Inc., 188 Asbury Street, S. Hamilton, MA 01982—*bulbs*

Fantastic Gardens, 9550 S W 67th Street, S. Miami, FL 33143—*tropical plants, bromeliads*

Farmer Seed & Nursery Co., Fairbault, MN 55021—*vegetables, herbs, fruits & supplies*

Fennell Orchid Co., 26715 S W 157th Street, Homestead, FL 33030—*orchids*

Architectural Fiberglass and Pottery Co., 2020 S. Robertson Blvd., Los Angeles, CA 90034—*architectural containers*

Henry Field Seed & Nursery Co., Shenandoah, IA 51601—*vegetables, dwarf fruit, herbs*

Fischer Greenhouses, Linwood, NJ 08221—*African violets, gesneriads*

Floralite Co., 4124 E. Oakwood Road, Oak Creek, WI 53154—*light fixtures, equipment & supplies*

Garden Way Living Center, 1186 Williston Rd., S. Burlington, VT 05401—*garden books and hard-to-find tools & supplies*

Gloeckner Co., Inc., 15 E. 26th Street, New York, NY 10010—*seeds, auto-waterers, supplies, containers*

Green Acres Nursery, 14451 Northeast Second St., N. Miami, FL 33161—*palms*

Griffin Greenhouse Supplies, Inc., 1619 Main Street, Tewksbury, MA 01876—*auto-waterers, supplies, equipment containers ("Durateck" planters)*

Gurney Seed & Nursery Co., Inc., Yankton, SD 57078—*vegetables, dwarf fruit, supplies*

Hauserman's Orchids, Box 363, Elmhurst, IL 60128—*orchids*

The House Plant Corner, Box 810, Oxford, MD 21654—*supplies, light units, equipment, plants*

Margaret Ilgenfritz, Box 665, Monroe, MI 48161—*orchids*

Johnson Cactus Gardens, Paramount, CA 90723—*cacti, succulents*

J. W. Jung Seed Co., Randolph, WI 53956—*vegetables, herbs, fruits*

Kartuz Greenhouses, 92 Chestnut Street, Wilmington, MA 01887—*house plants, bromeliads*

Kelly Bros. Nurseries, Inc., Dansville, NY 14437—*berries, herbs, dwarf fruit*

Kennebunk House, 43 Fletcher St., Kennebunk, ME 04043—*wick-type self-watering containers*

Krider Nurseries, Inc., Middlebury, IN 46540—*berries, dwarf fruit*

A. M. Leonard & Son, Inc., Piqua, OH 45356—*tools, equipment*

Logee's Greenhouses, 55 North Street, Danielson, CT 06239—*house plants*

Lyndon Lyon, 14 Mutchler St., Dolgeville, NY 13329—*African violets, columneas*

Manhattan Garden Supply, 305 N. Sepulveda Blvd., Manhattan Beach, CA 90266—*fuchsias, geraniums, supplies*

Earl May Seed & Nursery Co., Shenandoah, IA 51601—*vegetables, herbs, fruits*

Rod McLellan Co., 1450 El Camino Real, S. San Francisco, CA 94080—*orchids*

Mellingers, 2310 S. Range Rd., N. Lima, OH 44452—*seeds, plants, tubs, containers, tools, supplies*

Merry Gardens, Camden, ME 04843—*house plants*

The Meyer Seed Co., 600 S. Caroline Street, Baltimore, MD 21231—*vegetables, herbs, fruit, supplies*

H. F. Michells Co., King of Prussia, PA 19406—*seed, tubs & containers, supplies*

Midwest Seed Growers, 505 Walnut St., Kansas City, MO 64106—*vegetables*

Mini-Roses, Box 4255, Station A, Dallas, TX 75208—*miniature roses*

Musser Forests, Inc., P.O. Box 640, Indiana, PA 15701—*dwarf fruit*

Neas Growers Supply, P.O. Box 8543, Greenville, SC 29604—*table-top light units, equipment, supplies*

Neosho Nurseries, Neosho, MO 64850—*berries, dwarf fruit*

Nichols Garden Nursery, Albany, OR 97321—*imported European & Oriental vegetables*

Robert Nicholson Seed Co., Dallas, TX 75202—*vegetables, herbs*

Oakhurst Gardens, P.O. Box 444, Arcadia, CA 91008—*unusual bulbs, plants*

L. L. Olds Seed Co., P.O. Box 1069, Madison, WI 53701—*vegetables, supplies*

George W. Park Seed Co., Inc., Box 31, Greenwood, SC 29646—*vegetables, herbs, house plants, supplies*

Placet Products, 5000 W. 35th Street, Minneapolis, MN 55416—*decorative planters*

Plantamation, Inc., 136 E. 57th Street, New York, NY 10022—*self-watering containers*

A.L. Randall Co., 1325 W. Randolph Street, Chicago, IL 60607—*terrarium containers*

Rayner Brothers, Inc., Salisbury, MD 21801—*vegetables, dwarf fruit*

Julius Roehrs Co., East Rutherford, NJ 07973—*house plants*

Harvey J. Ridge, Co., 1126 Arthur Street, Wausau, WI 54401—*lights, fixtures, equipment & supplies*

Al Saffer & Co., Inc., 130 W. 28th Street, New York, NY 10001—*auto-waterers, containers, lights, equipment & supplies*

Harry Saier, Dimondale, MI 48821—*European & Oriental vegetable seeds*

Schell's Seed House, P.O. Box 2045, Harrisburg, PA 17105—*vegetables, herbs, fruits, supplies*

Seedway Inc., Hall, NY 14463—*Seeds*

Shep, 815 Main Street, Huntington Beach, CA 92646—*earthenware pots*

Shoplite. Co., 650 Franklin Avenue, Nutley, NJ 07110—*lights and equipment*

R. H. Shumway Seedsman, Rockford, IL 61101—*vegetables, fruit, supplies*

Stark Bro's Nurseries & Orchards, Louisiana, MO 63353—*dwarf fruit*

George Tait & Sons, Inc., 900 Tidewater Drive, Norfolk, VA 23516—*vegetables, supplies*

Tinari Greenhouses, 2325 Valley Rd., Bethayres, PA 19006—*African violets, other plants*

Tropical Paradise Greenhouses, 8825 W. 79th Street, Overland Park, KS 66200—*house plants, tropicals*

Tube Craft, Inc., 1311 W. 80th Street, Cleveland, OH 44102—*light equipment, supplies*

Verd-a-Ray Corp., 615 Front Street, Toledo, OH 43605—*lights and equipment*

Wetsel Seed Co., Inc., P.O. Box 791, Harrisonburg, VA 22801—*vegetables, herbs, supplies*

Wilson Bros., Roachdale, IN 46172—*geraniums, house plants*

Wyndham Hayward, 915 S. Lakemont Avenue, Winter Park, FL 32789—*tropical plants*

Some Selected References

A good library is essential for the person or family raising food and flowers. No one can remember all of the information this requires, and a good library will provide it, at your fingertips. New ideas, techniques and theories are always being put forth, and the best way to keep up with them all is to keep your library up to date. There are many good books available; here are some from Garden Way Publishing that are excellent choices.

Dwarf Fruit Trees for the Home Gardener, by Lawrence Southwick. 118 pp., quality paperback, $3.50; hardback, $5.95. All you need to know to start a home orchard.

The Complete Guide to Growing Berries and Grapes, by Louise Riotte. 142 pp., quality paperback, $3.95; hardback, $5.95. What to plant, where, when and exactly how.

Secrets of Companion Planting for Successful Gardening, by Louise Riotte. 225 pp., quality paperback, $4.95; hardback, $8.95. For bigger, more luscious crops. It works.

Vegetable Garden Handbook, by Roger Griffith. 120 pp., spiral bound $3.95. Take it into your garden, for information and your own record book.

Down-to-Earth Vegetable Gardening Know-How, featuring Dick Raymond. 160 pp., quality paperback, $4.95. Proven, practical advice for every level of vegetable gardener.

These books are available at your bookstore, or may be ordered directly from Garden Way Publishing, Dept. GFF, Charlotte, Vermont 05445. If your order is less than $10, please add 60¢ postage and handling.

Other books recommended by the author include the following:

The After-Dinner Gardening Book. Richard W. Langer. 1971. Collier Books, New York, N.Y.

Beginner's Guide to Hydroponics. James Sholto Douglas. 1972. Drake Publishers, Inc., New York, N.Y.

Bonsai For Beginners. H. J. Larkin. 1969. Arco Publishing Co., New York, N.Y.

The Complete Book of Houseplants. Charles Marden Fitch. 1972. Hawthorn Books Inc., New York, N.Y.

The Color Dictionary of Flowers and Plants for Home and Garden. Roy Hay and Patrick M. Synge. Crown Publishers, New York, N.Y.

The Complete Book of Gardening Under Lights. Elvin McDonald. Doubleday & Co., Inc., Garden City, N.Y.

Exotic Plant Manual. Alfred B. Graf. Roehrs Co., Book Division, E. Rutherford, N.J.

Gardening in Containers: A Handbook. 1963. Brooklyn Botanic Garden Record, Brooklyn, N.Y.

Gardens in Glass Containers. Robert C. Baur. 1970. Hearthside Press, Inc., New York, N.Y.

Herbs to Grow Indoors. Adelma Grenier Simmons. 1969. Hawthorn Books, Inc., New York, N.Y.

Indoor Gardens with Controlled Lighting. U.S. Dept. of Agric. 1971. Home and Garden Bul. No. 187. Supt. of Documents, U.S. Govt. Printing Office, Washington, D.C.

Minigardens for Vegetables. U.S. Dept. of Agric. 1970. Home and Garden Bul. No. 163. Supt. of Documents, U.S. Govt. Printing Office, Washington, D.C.

Modern Guide to House Plants. Ann Warren. 1961. New Era Books, Clinton, Ct.

Some Plants Adaptable to Container Culture

KEY TO SPECIAL NEEDS

Propagation:

- s — seed, in dark
- sl — seed, with light
- l — layering
- al — air layering
- rc — root cuttings
- swc — softwood cuttings
- hwc — hardwood cuttings
- lc — leaf cuttings
- div — divisions of crown
- tdv — tuber division
- cdr — corm division
- bdv — bulb division
- bd — budded

Watering:

- sd — soak, drain
- m — maintain moist
- w — maintain wet
- hh — high humidity

Light:

- fs — full sun, high light intensity, patio or south window
- dl — diffused light, easterly window
- st — shade tolerant, low light intensity, northerly window
- sil — supplemental indoor light
- SD — short day
- LD — long day

Other:

- drp — dormant rest period, dry rest period
- ctp — cool temperature preferred
- ntp — warm temperature preferred
- nwh — not winter hardy, not cold tolerant
- as — acid soil
- B — for bonsai use

NOTE: Where no symbols are given usually no special treatment or condition is required.

TREES—for Form and Foliage

Andromeda, Japanese *(Pieris japonica)* ctp, fs-dl, m, as

Arbor-vitae *(Thuja occidentalis)* m, fs-dl

Beech *(Fagus sylvatica)* m, fs

Cedar, blue Atlas *(Cedrus atlantica glauca)* m, fs

Deodar cedar *(C. deodara)* m, st, nwh

Fir, Nordmann *(Abies nordmanniana)* m, fs, wtp

Ginkgo *(Ginkgo biloba)* m, fs, B

Golden-chain tree *(Laburnum watereri)* nwh

Hinoki cypress *(Chamaecyparis obtusa)* m, fs, B

Juniper, Pfitzer *(Juniperus chinensis pfitzeriana)* hwc, dl, nwh

Juniper, prostrate *(J. communis depressa)* hwc, nwh, B

Juniper, Sargent *(J. chinensis sargentii)* hwc, B

Juniper, San Jose *(J. chinensis)* hwc, l, nwh, B

Laurel, Grecian *(Laurus nobilis)* m, fs, ctp, nwh, as

Loquat *(Eriobotrya japonica)* m, fs, nwh

Maple, Japanese *(Acer palmatum)* sd, drp, fs, nwh

Maple, trident *(A. buergerianum)* m, fs, nwh, B

Norfolk Island pine *(Araucaria excelsa)* s, swc, dl, nwh

Olive tree *(Olea europaea)* hwc, sd, nwh

Palm, clustering *(Chamaedorea seifrizii)* sd, st, wtp, nwh

Palm, clustering feather (*Ptychosperma* spp.) w, hh, wtp, nwh

Palm, Mediterranean fan *(Chamaerops humilis)* w, fs, nwh

Palm, sentry *(Howea belmoreana)* m, dl, nwh

Pine, bristle-cone *(Pinus aristata)* m, fs, B

Pine, Japanese black *(P. thunbergi)* fs, ctp, B

Pine, Japanese red *(P. densiflora)* fs, ctp, B

Pine, Japanese white *(P. parviflora glauca)* fs, ctp, B

Pine, lace-bark *(P. bungeana)* m, B

Pine, Mugo *(P. mugo mughus)* fs, ctp, B

Pine, Norway *(P. resinosa)* fs, ctp, B

Rubber plant *(Ficus elastica)* al, m, dl, nwh

Spruce, Colorado blue *(Picea pungens glauca)* m, fs, B

Spruce, dwarf white *(P. alba conica)* m, dl, nwh

Spruce, Norway *(P. excelsa)* sd, fs, wtp, nwh

Umbrella pine *(Sciadopitys verticillata)* al, nwh

Umbrella tree *(Schefflera actinophylla)* m, B

Weeping fig *(Ficus benjamina) al, nwh*

Zelkova *(Zelkova serrata) m, b*

TREES—for Flower and Fruit

Apple *(Pyrus malus)* sil, drp

Apricot *(Prunus armeniaca)* drp, nwh

Avocado *(Persea americana)* fs, st, nwh

Calamondin orange *(Citrus mitis)* sd, nwh, B

Cherry (*Prunus* spp.) drp, nwh, B

Cherry, Japanese flowering *(P. serrulata)* fs, nwh, B

Coffee *(Coffea arabica)* as, wt, nwh

Crabapple, flowering *(Malus floribunda)* drp

Cranberry tree, dwarf European *(Viburnum opulus roseum)* ctp, nwh

Crape myrtle *(Lagerstroemia indica)* fs, nwh

Dogwood, flowering *(Cornus florida rubra)* nwh

Fig *(Ficus carica)* al, nwh

Flowering maple *(Abutilon hybridum)* hwc, nwh, B

Grapefruit *(Citrus paradisi)* sd, bd, nwh

Guava *(Feijoa sellowiana)* sd, nwh

Hawthorn, Paul's scarlet *(Crataegus monogyna pauli)* sd, nwh

Holly, American *(Ilex opaca)* ctp, fs, m, as

Holly, Chinese *(I. cornuta)* fs, m, nwh, as

Holly, English *(I. aquifolium)* fs, m, nwh, as

Holly, Japanese *(I. crenata)* m, nwh, B

Kumquat, Nagami *(Fortunella margarita)* sl, m, nwh, B

Lemon *(Citrus limonia)* sd, bd, nwh, B

Lime *(C. aurantifolia)* sd, bd, nwh, B

Magnolia *(Magnolia grandiflora)* l, nwh

Mulberry *(Morus rubra)* fs, drp

Nectarine *(Prunus persica nectarina)* nwh

Orange *(Citrus sinensis)* bd, sd, sil, nwh, B

Palm, dwarf coconut *(Cocos nucifera)* hh, wtp, nwh

Palm, large lady *(Rhapsis excelsa)* w, nwh

Palm, pigmy date *(Phoenix loueiri dactylifera)* s, wtp, w, nwh

Papaya *(Carica Papaya)* s, m, fs, nwh

Pawpaw, South American *(Carica quercifolia)* s, m, fs, nwh

Peach *(Prunus persica)* sil, hwc, nwh, drp

Pear, dwarf *(Margyricarpus setosus)* l, drp, nwh

Persimmon *(Diospyros virginiana)* nwh

Plum, flowering *(Prunus blireana)* nwh, drp

Pomegranate, dwarf *(Punica granatum nana)* fs, nwh, B

Sweet olive *(Osmanthus fragrans)* fs, nwh

Tangelo (*Citrus* hybrid) sd, sl, fs, nwh

Tangerine *(C. nobilis deliciosa)* bd, sil, nwh

SHRUBS—for Form and Foliage

Aralia, Japanese *(Fatsia japonica)* swc, st, nwh

Barberry *(Berberis julianae)* hwc, m, fs

Boxwood (*Buxus* spp.) hwc, nwh

Camellia *(Camellia sasanqua)* swc, sl, nwh, B

English laurel *(Prunus laurocerasus)* nwh

Fern pine *(Podocarpus gracilior)* nwh

Firethorn *(Pyracantha coccinea lalandii)* swc, nwh

Gold-dust tree *(Aucuba japonica)* sd, nwh

Holly olive *(Osmanthus ilicifolius variegatus)* ctp, fs, nwh

Japanese privet *(Ligustrum japonicum)* sd, sl, nwh

Mahonia *(Mahonia lomariifolia)* sl, nwh

Mountain laurel *(Kalmia latifolia)* as, st, nwh

Myrtle, Greek *(Myrtus communis)* sl, swc, nwh, sd

Natal plum *(Carissa grandiflora)* nwh

Poinciana, dwarf *(Poinciana pulcherrima)* sl, nwh

Schefflera (*Schefflera* spp.) sl, nwh, sd

Sea-grape *(Coccolobis uvifera)* sl, fs, nwh

Silver berry *(Elaeagnus pungens aureo/variegata)* fs, nwh

Sweetbay Magnolia *(Magnolia virginiania)* m, hwc, drp, ctp, LD

Tea plant *(Camellia sinensis)* sl, nwh

Viburnum "Laurestinus" *(Viburnum tinus)* m, sl, nwh

Yew pine *(Podocarpus macrophyllas)* sl, nwh

SHRUBS—for Flower and Fruit

Abelia, glossy *(Abelia grandiflora)* st, nwh

Angel's trumpet (*Datura arborea* and *D. suaveolens)* nwh

Azalea (*Azalea* spp. esp. Ghent and *mollis* hybrids) l, swc, sl, nwh, as

Azalea, Kurume (*Azalea* hybrid) as, sl, nwh, B, as

Banana, dwarf *(Musa nana)* sl, swc, nwh

Barberry, Japanese *(Berberis thunbergii)* d, m, fs

Blackberry *(Rubus* spp.) rc, fs

Blueberry, highbush *(Vaccinium corymbosum)* as, fs

Bougainvillea (*Bougainvillea* spp.) sl, swc, sd, nwh

Bouvardia *(Bouvardia ternifolia)* sl, nwh, drp

Brunfelsia *(Brunfelsia calycina)* dl, nwh

Butterfly bush *(Buddleja asiatica)* nwh

Camellia *(Camellia japonica)* swc, dl, sl, nwh

Carissa *(Carissa grandiflora nana)* nwh

Ceanothus *(Ceanothus delilianus Gloire des Versailles)* swc, nwh

Cotoneaster *(Cotoneaster horizontalis)* l, s, B, nwh

Currant (*Ribes* spp.) swc, m, dl

Daphne *(Daphne odora)* swc, l, dl, nwh

Deutzia (*Deutzia gracilis* hybrids) hwc, st, nwh

Elderberry *(Sambucus canadensis)* hwc, w, fs

Forsythia, Korean *(Forsythia ovata)* l, hwc, sd, nwh

Gardenia *(Gardenia jasminoides)* swc, sl, wtp, nwh

Gooseberry (*Ribes* spp.) l, dl-st

Hibiscus, Chinese *(Hibiscus rosa-sinensis)* swc, wtp, fs, nwh

Honeysuckle, bush *(Diervilla lonicera maackii)* swc, st, nwh

Hydrangea, azure (*Hydrangea macrophylla* "Enziandom") swc, as, drp

Hydrangea, carmine (*H. macrophylla* "Chapeau Rouge") swc, as, drp

Hydrangea, oak-leaf *(H. quercifolia)* swc, as, drp

Inkberry *(Ilex glabra compacta)* st, nwh

Japanese Pittosporum *(Pittosporum tobira)* fs, drp, nwh

Japanese Quince, flowering *(Chaen-*

omeles lagenaria) swc, fs, hwc, l, nwh, B
Leadwort *(Plumbago auriculatum)* fs, wtp
Lilac (*Syringa* spp.) l, swc, drp
Lily-of-the-valley shrub *(Pieris japonica)* ctp, dl
Mango *(Mangifera indica)*, s, wtp, nwh
Monkey flower *(Mimulus aurantiacus)* s, swc, nwh
Necklace bush *(Enkianthus cernuus)* nwh
Oleander *(Nerium oleander)* swc, st, nwh
Oregon grape *(Mahonia aquifolium)* fs, drp, nwh
Poinsettia *(Euphorbia pulcherrima)* swc, wtp, sd, sl, nwh
Pyracantha or Firethorn *(Pyracantha coccinea)* swc, fs, st, m
Quince *(Cydonia oblonga)* l, nwh
Raspberry (*Rubus* spp.) rc, fs
Rhododendron, Carolina *(Rhododendron carolinianum)* swc, m, as, nwh
Rhododendron, Catawba *(Rhododendron catawbiense)* swc, m, as, nwh
Rose (*Rosa* spp.) swc, nwh
Rose, miniature (*Rosa chinensis* varieties) sl, nwh
Rose-of-Sharon or Althea *(Hibiscus syriacus albus)* dl, nwh
Smoke Tree *(Cotinus coggygria)* m, nwh
Streptosolen jamesoni *(Streptosolen jamesonii)* fs, m, nwh
Yesterday today and tomorrow *(Brunfelsia calycina floribunda)* m, nwh
Yeddo Hawthorn *(Raphiolepis umbellata ovata)* nwh

VINES and CLIMBERS

Allamanda *(Allamanda cathartica)* fs, nwh
Allegheny vine *(Adlumia fungosa)* st, nwh
Black-eyed Susan vine *(Thunbergia alata)* m, st, hh, nwh
Canary bird flower *(Tropaeolum peregrinum)* s, st, nwh
Cape leadwort *(Plumbago capensis)* swc, fs, nwh
Clematis, marsh *(Clematis crispa)* st, nwh
Clematis, scarlet *(C. texensis)* st, nwh
Clematis, "Sweet Autumn" *(C. paniculata)* st, nwh
Columbine (*Aquilegia* spp.) d, fs
Cup and saucer vine *(Cobaea scandens)* s, fs, nwh
Cup of gold *(Solandra guttata)* fs, nwh
Fatshedera *(Fatshedera lizei)* fs-st, nwh
Gloryosa lily *(Gloriosa rothschildiana)* d, wtp, nwh, drp
Gourd (several genera) fs, nwh
Grape (*Vitis* spp.) hwc, drp
Hop, Japanese *(Humulus japonicus)* sd, fs, nwh
Ivy, English *(Hedera helix)* swc, l, m, fs
Ivy, Japanese *(H. helix gracilis)* swc, l, m, fs
Jasmine (*Jasminum* spp.) sl, l, nwh
Kangaroo vine *(Cissus antarctica)* fs, sd, nwh
Madegascar jasmine *(Stephanotis floribunda)* wtp, fs, nwh
Moonflower *(Calonyction aculeatum)* s, nwh
Morning glory *(Ipomoea purpurea)* s, nwh
Passion flower (*Passiflora* spp.) swc, nwh
Pepper plant *(Piper nigrum)* st, wtp, nwh
Potato vine *(Solanum jasminoides)* wtp, nwh
Purple passion vine *(Gynura sarmentosa)* m, nwh
Scarlet Kadsura *(Kadsura japonica)* fs, nwh
Scarlet runner bean *(Phaseolus coccineus)* fs, nwh
Silver lace vine *(Polygonum aubertii)* hh, st, nwh
Trumpet vine *(Campsis radicans)* fs, st, drp
Vinca vine *(Vinca major)* s, nwh
Winter creeper *(Euonymus radicans)* fs-dl, nwh
Wisteria *(Wisteria floribunda macrobotrys)* swc, m, fs, drp

PLANTS for Hanging Baskets

Achimenes *(Achimenes longiflora alba)* swc, dl, hh, m, nwh

Asparagus fern *(Asparagus plumosus)* div, nwh

Begonia, trailing *(Begonia limmingheiana)* rc, m, wtp, st, nwh

Bindweed *(Convolvulus mauritanicus)* swc, fs-dl, nwh

Blue pimpernel *(Anagallis linifolia)* s, fs, nwh

Browallia *(Browallia speciosa Major)* sl, st, wtp, nwh

Campanula, trailing *(Campanula fragilis)* s, m. fs, nwh

Clock-vine *(Thunbergia grandiflora)* fs-dl, nwh

Coleus, trailing *(Coleus rehneltianus)* sl, swc, wtp, nwh

Dianthus *(Dianthus plumarius)* l, fs, nwh

Emerald Fern *(Asparagus sprengeri)* tdv, hh, wtp, nwh

Flame violet *(Episcia cupreata)* wtp, hh, fs-dl, nwh

Flowering inch plant *(Tradescantia blossfeldiana)* fs, nwh

Fuchsia, trailing *(Fuchsia procumbens)* swc, st, nwh

Ivy, English *(Hedera helix)* hwc, l, ctp, swc, fs-st

Ivy, Kenilworth *(Cymbalaria muralis)* s, swc, m, nwh

Ivy-leaved geranium *(Pelargonium peltatum)* ctp, fs, sd, nwh, drp

Lantana, trailing *(Lantana montevidensis)* ctp, fs, sd, nwh, drp

Lobelia *(Lobelia erinus pendula)* s, fs, st, m, nwh

Nasturtium, climbing *(Tropaeolum majus)* s, swc, ctp, nwh

Petunia, cascading *(Petunia hybrida)* sl, fs, st, m, nwh

Rose-moss *(Portulaca grandiflora)* s, wtp, fs, sd, nwh

Sedum *(Sedum spurium, S. hybridum, S. sarmentosum)* lc, sd, ctp, fs, nwh

Sedum, donkey's tail *(Sedum morganianum)* lc, fs-dl, sd, nwh

Shrimp plant *(Beloperone guttata)* swc, st, nwh, drp

Squirrel's foot fern *(Davallia bullata)* div, m, fs-dl, nwh, as

Sweet alyssum *(Alyssum maritimum)* s, m, fs, nwh

Tradescantia *(Tradescantia* spp.) swc, m, dl, nwh, drp

Tuberous begonia *(Begonia x tuberhybrida pendula)* s, swc, tdv, st, nwh

Wishbone flower *(Torenia fournieri)* s, dl, m, nwh

Zebrina *(Zebrina pendula)* dl, m, nwh

GROUND COVERS for Horizontal Emphasis

Bigleaf Wintercreeper *(Euonymus fortunei coloratus)* st, nwh

Carpet bugleweed *(Ajuga reptans)* fs-st, nwh

Coral bead plant *(Nertera depressa)* fs, sd, nwh

Creeping Jennie *(Lysimachia nummularia)* s, div, swc, fs-dl

Geranium ivy *(Parthenocissus tricuspidata Lowi)* wtp, m, dl, nwh

Ground Ivy *(Glecoma hederacea)* m, fs-st

Ivy, German *(Senecio mikanioides)* swc, dl, SD, nwh

Kalanchoe pumila *(Kalanchoe pumila)* s, lc, swc, fs, sd, drp

Moneywort *(Lysimachia nummularia)* fs, m

Pachysandra *(Pachysandra terminalis)* st

Periwinkle (Creeping Myrtle) *(Vinca minor)* swc, m, fs-st

Purple Heart *(Setcreasea purpurea)* wtp, drp, nwh

Rattlesnake *(Crassula feres)* s, lc, swc, fs, sd, drp

Tufted Sandwort *(Arenaria grandiflora)* st, nwh

ANNUAL FLOWERS

African daisy *(Arctotis grandis)* s, fs, drp

Ageratum *(Ageratum houstonianum)* s, sl, fs

Balsam *(Impatiens balsamina)* m, st, swc

Calendula *(Calendula officinalis)* s, m, ctp

Candytuft *(Iberis umbellata)* s, fs

Cape marigold *(Dimorphotheca au-*

rantiaca) s, fs
Christmas Cherry *(Solanum pseudocapsicum)* s, fs, nwh
Christmas pepper *(Capsicum frutescens)* s, fs, nwh
Cockscomb *(Celosia argentea cristata)* s, SD, wtp
Coleus *(Coleus blumei)* sl, swc, fs-st, nwh
English daisy *(Bellis perennis)* s, fs drp
Gloriosa daisy *(Rudbekia hirta)* s, fs, ld, drp
Marigold , French *(Tagetes patula)* s, LD, fs
Nemesia *(Nemesia strumosa)* s, dl, ctp
Nicotiana *(Nicotiana alata grandiflora)* sl, fs, m
Pansy *(Viola tricolor hortensis)* s, wtp, fs
Pepper, ornamental *(Solanum capsicastrum)* s, wtp, fs, sl
Petunia *(Petunia hybrida)* sl, fs-st, wtp
Phacelia *(Phacelia campanularia)* fs, m
Phlox *(Phlox drummondii)* s, LD, fs
Primrose *(Primula malacoides)* sl, SD, wtp
Salvia *(Salvia farinacea)* sl, swc, fs, SD
Scabiosa *(Scabiosa atropurpurea)* s, LD, fs, m
Stock *(Mathiola incana)* s, m, fs
Snapdragon (*Antirrhinum* spp.) sl, fs, LD
Sweet pea *(Lathyrus odoratus)* s, ctp, fs
Torenia *(Torenia fournieri)* s, dl-st
Verbena (*Verbena* spp.) fs
Viper's Bugloss *(Echium lycopsis)* fs
Viscaria *(Lychnis coeli-rosa)* fs
Zinnia (*Zinnia* spp.) s, SD, fs, m

BIENNIAL and PERENNIAL FLOWERS

African violet *(Saintpaulia ionantha)* lc, wtp, sil, hh, nwh
Aster *(Callistephus chinensis)* div, swc, fs
Bears breach *(Acanthus mollis)* s, rc, div, fs, dl, nwh
Begonia (*Begonia* spp.) lc, sl, st, nwh
Begonia, miniature *(B. aridicaulis)* lc, sl, st, nwh
Bellflower *(Campanula isophylla)* m, dl
Bermuda buttercup *(Oxalis pes-cernua)* fs, nwh
Bird of paradise *(Strelitzia reginae)* fs, nwh, drp
Canna, dwarf *(Canna generalis Pfitzeri)* tdv, fs, nwh, drp
Carnation (*Dianthus* spp.) s, ctp, fs, nwh, drp
Christmas rose *(Helleborus niger)* ctp, drp
Chrysanthemum (*Chrysanthemum* spp.) swc, fs
Cineraria *(Senecio cineraria)* s, st, fs-dl, nwh, drp
Coreopsis (*Coreopsis* spp.) fs, LD, nwh
Cup flower *(Nierembergia caerulea)* sd, m, nwh
Cyclamen, florist's *(Cyclamen persicum)* s, tdv, dl, hh, nwh
Dimorphotheca *(Dimorphotheca aurantiaca)* swc, ctp, fs-dl, nwh
Egyptian star cluster *(Pentas lanceolata)* wtp, fs, nwh
Exacum *(Exacum affine)* wtp, dl, nwh
Flamingo flower *(Anthurium scherzerianum)* wtp, w, hh, st, nwh
Forget-me-not *(Myosotis alpestris)* s, fs, drp, as
Foxglove *(Digitalis purpurea)* div, m, fs, drp, as
Geranium *(Pelargonium* spp.) swc, fs, nwh, drp
Gloxinia *(Sinningia speciosa)* sl, lc, st, sd, nwh, drp
Grecian urn plant *(Acanthus mollis)* dl, m, nwh
Heliotrope *(Heliotropium aborescens)* s, swc, st, wtp, nwh
Hollyhock *(Althaea rosea)* fs-st, drp
Iris (*Iris* spp.) div, fs, drp
Lantana *(Lantana camara)* wtp, fs, nwh, drp
Lenten-rose *(Helleborus orientalis atrorubens)* m, fs, drp
Leucothoë, drooping *(Leucothoë catesbaei)* ctp, fs, nwh
Lily of the Nile *(Agapanthus africanus)* dl, nwh, drp

Lily-of-the-valley *(Convallaria majalis)* div, drp

Mexican foxglove *(Allophytan mexicanum)* m, hh, dl, nwh

Orchid, florist's *(Cattleya* spp.) sl, w, st, hh, nwh

Orchid, lady's slipper *(Cypripedium* spp.) sl, w, st, hh, nwh

Orchid, miniature *(Epidendrum* spp.) sl, w, st, hh, nwh

Orchid, dwarf rainbow *(Ascocentrum miniatum)* sl, sd, dl, hh, nwh

Orchid, spray *(Oncidium* spp.) sl, w, st, hh, nwh

Orchid, tiger *(Odontoglossum* spp.) sl, w, st, hh, nwh

Pikake Flower m, fs, dl, nwh

Pink *(Dianthus* spp.) fs, drp

Plantain lily *(Hosta* spp.) m, dl, ctp, drp

Ranunculus (Persian buttercup) *(Ranunculus asiaticus)* tdv, fs-dl, drp

Rex begonia *(Begonia rex)* lc, wtp, m, hh, dl, nwh

Rose-mallow *(Hibiscus moscheutos)* m, fs, nwh, drp

Sundrop, tufted *(Oenothera caespitosa)* m, fs

Star jasmine *(Trachelospermum jasminoides)* swc, l, fs-st, nwh, drp

Sweet William *(Dianthus barbatus)* m, fs

Trailing arbutus *(Epigaea repens)* ctp, dl-st, as

Transvaal daisy *(Gerberia jamesonii flore-pleno)* ctp, sd, fs, nwh

Tree-marguerite *(Chrysanthemum frutescens)* swc, ctp, m, fs, nwh

Veronica or Speedwell *(Veronica spicata)* ctp, fs, nwh

FLOWERING BULBS, CORMS, TUBERS, etc.

Amaryllis *(Amaryllis vittata)* s, bdv, fs-dl, m, wtp, nwh, drp

Blood lily *(Haemanthus katherinae)* bdv, fs, m, nwh, drp

Calla, golden *(Zantedeschia elliottiana)* bdv, wtp, fs, w, nwh, drp

Crocus *(Crocus* spp.) cdv, ctp, drp

Daffodil *(Narcissus* spp.) bdv, fs, m, drp

Dahlia *(Dahlia* spp.) tdv, fs-dl, m, nwh, drp

Elephant's ear *(Colocasia esculenta)* tdv, dl, hh, nwh, drp

Freesia *(Freesia hybrida)* s, cdv, fs, sd, nwh, drp

Gladiolus *(Gladiolus* spp.) cdv, fs, nwh, drp

Grape-hyacinth *(Muscari* spp.) bdv, ctp, fs, w, drp

Hyacinth *(Hyacinthus orientalis)* bdv, ctp, fs-dl, w, drp

Kafir lily *(Clivia miniata)* tdv, fs-dl, sd, nwh

Lily, dwarf orange *(Lilium elegans)* bdv, sd, fs, drp

Lily, dwarf pink *(L. rubellum)* bdv, sd, fs, drp

Lily, dwarf scarlet *(L. tenuifolium)* bdv, sd, fs, drp

Lily, dwarf white *(L. philippinense)* bdv, sd, fs, drp

Narcissus *(Narcissus* spp.) bdv, fs, drp

Nerine *(Nerine curvifolia)* bdv, fs, nwh, drp

Scilla *(Scilla* spp.) bdv, fs-dl, drp

Snowdrop *(Galanthus* spp.) bdv, ctp, dl, drp

Spider lily *(Hymenocallis speciosa)* bdv, wtp, fs, nwh, drp

Squill, Siberian *(Scilla sibirica)* bdv, fs, drp

Tiger flower *(Tigridia pavonia)* bdv, fs, m, nwh, drp

Tulip *(Tulipa* spp.) bdv, ctp, fs, drp

Zephyr-lily *(Zephyranthes grandiflora)* bdv, ctp, m, nwh, drp

FOLIAR PLANTS for Patio and House

Adam's Needle *(Yucca filamentosa)* div, fs, sd, nwh

Aechmea "Silver King" *(Aechmea fasciata)* wtp, dl, m, nwh

Aeonium *(Aeonium* spp.) lc, sd, nwh

Agave *(Agave attenuata)* lc, fs, nwh

Aloe *(Aloe* spp.) lc, fs-dl, nwh

Anthurium *(Anthurium* spp.) wtp, hh, w, st, nwh

Aphelandra *(Aphelandra squarrosa)* wtp, dl, nwh

Apostle plant *(Neomarica northiana)* div, dl, nwh

Bamboo, black *(Phyllostachys nigra)* div, fs, m, nwh

Bamboo, Chinese Goddess *(Bambusa multiplex riviereorum)* div, fs, m, nwh

Bamboo, dwarf *(Arundinaria japonica)* div, fs, m, nwh

Bamboo, golden *(Phyllostachys aurea)* div, fs, m, nwh

Bamboo, hardy *(P. viridis)* div, fs, nwh

Bamboo, heavenly *(Nandina domestica)* div, fs, m, nwh

Bamboo, palmate *(Sasa palmata)* div, fs, nwh

Bishop's Cap *(Astrophytum myriostigma)* wtp, sd, fs-dl-nwh

Bromeliads (many genera) swc, fs, nwh

Broom-corn *(Holcus sorghum technicus)* s, wtp, nwh

Cactus, Crab *(Zyocactus truncata)* swc, wtp, sd, dl, nwh

Cactus, Christmas *(Schlumbergera bridgesii)* swc, wtp, m, dl, nwh

Cactus, Easter lily *(Echinopsis multiplex)* swc, wtp, sd, fs, nwh

Cactus, orchid (*Epiphyllum* hybrids) swc, sd, dl, nwh

Cactus, Peruvian old man *(Espostoa lanata)* swc, sd, fs, nwh

Cactus, prickly pear (*Opuntia* spp.) swc, sd, fs, wtp, nwh

Cactus, sand dollar or sea urchin *(Astrophytum asterias)* swc, sd, fs, wtp, nwh

Caladium *(Caladium bicolor)* tdv, wtp, fs-dl, nwh, drp

Cast iron plant *(Aspidistra elatior)* st, nwh

Castor bean *(Ricinus communis)* s, fs, nwh

Century plant *(Agave americana marginata)* lc, fs-st, nwh

Chinese evergreen *(Aglaonema modestum)* w, dl-st, nwh

Corn plant *(Dracaena fragrans massangeana)* lc, dl, nwh

Crassula *(Crassula falcata)* lc, fs-dl, sd, nwh

Crown of thorns *(Euphorbia splendens)* lc, wtp, fs, sd, nwh

Cryptanthus zonatus *(Cryptanthus zonatus)* s, dl, nwh

Decorator plant *(Dracaena marginata)* swc, wtp, dl, w, nwh

Dracaena deremensis *(D. deremensis)* wtp, dl-st, nwh

Dudleya (*Dudleya* spp.) lc, fs-dl, nwh

Dumbcane (*Dieffenbachia* spp.) swc, dl, sd, nwh

Earth stars *(Cryptanthus acaulis)* wtp, dl, sd, nwh

Felt-bush *(Kalanchoe beharensis)* sl, lc, sd, fs, nwh

Fern, asparagus *(Asparagus plumosus)* div, sd, m, dl-st, nwh, as

Fern, birdsnest *(Asplenium nidus)* swc, w, st, nwh, as

Fern, Boston *(Nephrolepis exaltata bostoniensis)* div, sd, hh, dl-st, nwh, as

Fern, Deer Tongue *(Blechnum spicant)* div, sd, nwh, as

Fern, five-finger *(Adiantum pedatum)* div, wtp, w, hh, st, nwh, as

Fern, lace *(Cheilanthes gracillima)* div, sd, nwh, as

Fern, Lady *(Athyrium filix-femina)* div, sd, as

Fern, maidenhair (*Adiantum* spp.) div, wtp, w, hh, st, as

Fern, round-leaf *(Pellaea rotundifolia)* div, m, hh, st, nwh, as

Fern, sword *(Nephrolepsis exaltata)* div, m, fs-dl, nwh, as

Fiddle leaf Fig *(Ficus lyrata)* swc, al, m, fs-dl, nwh

Flax, New Zealand *(Phormium tenax atropurpureum)* fs, m, nwh

Flowering Cabbage *(Brassica oleracea acephala)* s, ctp, fs, m, nwh

German ivy *(Senecio mikanioides)* swc, sd, wtp, fs-dl, nwh

Haworthia Tessellata *(Haworthia tessellata)* lc, sd, dl, sd, nwh

Hen-and-chickens *(Sempervivum tectorum)* lc, ctp, fs, sd

Ice plant *(Mesembryanthemum crystallinum)* lc, fs, sd, nwh

Jade plant *(Crassula argentea)* s, fs-dl, nwh

Mexican Snowball *(Echeveria elegans)* lc, fs, sd, nwh

Mimosa or Sensitive plant *(Mimosa pudica)* s, hh, fs-dl, nwh

Oyster plant *(Rhoeo discolor vittata)* s, fs-dl, nwh

Panda plant *(Kalanchoe tomentosa)* sl, lc, fs, sd, nwh

Parlor palm *(Chamaedorea elegans)* wtp, fs-st, m, nwh

Peperomia (*Peperomia* spp.) s, dl, nwh

Philodendron *(Philodendron sellorum)* swc, dl, w, nwh

Philodendron *(P. oxycardium)* swc, dl, w, nwh

Philodendron, finger-leaf *(P. bipinnatifidum)* swc, dl, w, nwh

Philodendron, split-leaf *(Monstera deliciosa)* swc, dl, w, nwh

Piggy-back plant *(Tolmiea menziesi)* wtp, fs-dl, m, nwh

Prayer plant *(Maranta leuconeura kerchoveana)* lc, hh, m, dl, nwh

Rubber plant *(Ficus elastica variegata)* lc, al, wtp, fs-dl, m, nwh

Sedum (*Sedum* spp.) lc, sd, ctp, fs, nwh

Screw pine *(Pandanus veitchii Compactus)* s, fs-dl, nwh

Slipperwort (*Calceolaria herbeohybrida* "Grandiflora") s, ctp, fs-st, sd, nwh

Spanish bayonet (*Yucca aloifolia* "Marginata") fs, sd, nwh

Snake plant *(Sansevieria trifasciata)* rc, fs, sd, nwh

Stone crop *(Crassula schmidtii)* lc, fs-dl, nwh

Thread-leaf false aralia *(Dizygotheca elegantissima)* swc, wtp, fs-dl, nwh

Umbrella plant *(Cyperus alternifolius)* dl, w, nwh

PLANTS for Pool and Fountain

Arrowhead *(Sagittaria engelmanniana)* w, fs, drp

Cat-tail *(Typha latifolia)* w, fs, drp

Chinese Evergreen *(Aglaonema modestum)* w, dl-st, nwh

Horsetail *(Equisetum hyemale)* w, fs-st

Jack-in-the-pulpit *(Arisaema triphyllum)* w, hh, fs-dl, drp, as

Japanese iris *(Iris kaempferi)* rc, w, fs, drp

Lotus, East Indian *(Nelumbium nelumbo)* w, fs, nwh, drp

Mint (*Mentha* spp.) s, swc, m-w, fs

Papyrus, dwarf *(Cyperus haspan viviparus)* w, dl, nwh

Rice, wild *(Zizania aquatica)* w, fs, nwh

Rush, horsetail *(Equisetum arvense)* m-w, fs-st

Water-hyacinth *(Eichornia crassipes)* wtp, hh, w, dl, nwh

Water lily, Pigmy *(Nymphaea tetragonahelvola)* wtp, fs-dl, w, drp

Water Snowflake *(Nymphoides indicum)* w, fs, nwh

Water cress *(Nasturtium officinale)* sl, w, fs, sil

VEGETABLES

Bean (*Phaseolus* spp.) fs, wtp, m

Beet *(Beta vulgaris)* m, fs

Broccoli *(Brassica oleracea italica)* m, fs, ctp

Brussels Sprouts *(B. oleracea gemmifera)* m, fs, ctp

Cabbage *(B. oleracea capitata)* fs, m

Carrot *(Daucus carota sativa)* m, fs, sil

Cauliflower *(Brassica oleracea botrytis)* m, fs, wtp

Corn, Indian *(Zea mays)* fs, wtp

Corn, pop *(Z. mays everta)* fs, wtp

Corn, sweet *(Z. mays saccharata)* m, fs, wtp

Cucumber *(Cucumis sativus)* fs, m, wtp

Eggplant *(Solanum melongena esculentum)* s, fs, wtp

Endive *(Cicnorium endivia)* fs, ctp

Kohlrabi *(Brassica caulorapa)* m, ctp, fs

Leek *(Allium porrum)* m, ctp, fs

Lettuce *(Lactuca sativa)* m, ctp, fs, sil

Melon *(Cucumis melo)* fs, m, wtp

Onion *(Allium cepa)* m, ctp, fs-dl, drp

Peanut, Spanish *(Arachis hypogaea)* m, fs, wtp, nwh, drp

Pepper *(Capsicum fruticosum)* m, wtp, fs, sil

Pineapple, miniature *(Ananas nanus)* m, hh, wtp, fs

Potato *(Solanum tuberosum)* tdv, m, ctp, fs, as

Potato, sweet *(Ipomoea batatas)* rc, m, wtp, fs

Radish *(Raphanus sativus)* ctp, fs, sil

Soybean *(Glycine max)* s, fs, wtp

Spinach *(Spinacia oleracea)* m, ctp, fs

Spinach, New Zealand *(Tetragonia expansa)* m, ctp, fs-st
Squash (*Cucurbita* spp.) m, fs, wtp
Swiss chard *(Beta vulgaris cicla)* m, ctp, fs
Tampala *(Amaranthus gangeticus)* m, wtp, fs
Tomato, midget *(Lycopersicon esculentum)* m, wtp, fs

HERBS

Basil *(Ocimum minimum)* s, swc, ctp, fs
Bay, sweet *(Laurus nobilis)* m, ctp, fs-dl, nwh
Borage *(Borago officinalis)* s, ctp, fs, sil
Chervil *(Anthriscus cerefolium)* s, ctp, fs-dl, sil
Chives *(Allium schoenoprasum)* s, ctp, fs-dl, sil
Dill *(Anethum graveolens)* s, fs, sil
Fennel, sweet *(Foeniculum vulgare dulce)* s, fs, ctp
Lavender *(Lavandula officinalis)* s, fs, sil
Lemon balm *(Melissa officinalis)* s, swc, sd, fs, sil
Marjoram, sweet *(Origanum majorana hortensis)* s, fs, sil
Mint (*Mentha* spp.) s, swc, m-w, fs, sil
Oregano *(Origanum vulgare)* s, fs, sd
Parsley *(Petroselinum crispum)* s, ctp, fs-dl, sil
Rosemary *(Rosmarinus officinalis)* s, swc, l, fs, sd
Sage *(Salvia officinalis)* s, swc, fs, sil
Savory, summer *(Satureja hortensis)* sl, swc, fs
Tarragon *(Artemisia dracunculus)* swc, d, st
Thyme (*Thymus* spp.) s, swc, wtp, fs
Thyme, lemon *(T. citriodorus)* s, wtp, fs

PLANTS for Bottle and Minigardens

Adiantum fern *(Adiantum hispidulum)* wtp, hh, st, as
Alternathera versicolor *(Alternathera versicolor)* wtp, fs, m
Baby's tears *(Helxine soleirolii)* m, dl, hh
Calanthea *(Calanthea micans)* wtp, hh, dl
Calico hearts *(Andromischus maculatus)* fs-dl, sd
Checkerberry or Wintergreen *(Gaultheria procumbens)* w, dl, as
Cinquefoil, creeping *(Potentilla nitida)* ctp, sd, fs-dl
Climbing fern *(Lygodium japonicum)* m, hh, dl, as
Club moss (*Lycopodium* spp.) ctp, dl, m, as
Cobweb houseleek *(Sempervivum arachnoideum)* ctp, fs, sd
Cyprus, dwarf false *(Chamaecyparis obtusa compacta)* ctp, fs, m
Dwarf club moss *(Selaginella kraussiana browni)* m, hh, st
Dyckia fosteriana *(Dyckia fosteriana)* dl, sd
Euonymus, "Silver Queen" *(Euonymus japonicus argenteo-variegatus)* ctp, m, dl
Euonymus, dwarf winged *(Euonymus alatus compactus)* ctp, m, dl
Eyelash begonia *(Begonia bowerii)* wtp, m, hh, dl
Gasteria liliputana *(Gasteria liliputana)* fs-dl
Glory-of-the-snow *(Chionodoxa luciliae)* ctp, fs, m, drp
Gloxinia, miniature *(Sinningia pusilla)* lc, wtp, dl
Grape-ivy, miniature *(Cissus striata)* fs-dl, m
Inch-plant *(Tradescantia multiflora)* fs-dl, sd
Jewel orchid *(Haemaria discolor dawsoniana)* wtp, sd, dl, as
Kalanchoe pumila *(Kalanchoe pumila)* s, lc, swc, fs, sd, drp
Leptotes bicolor *(Leptotes bicolor)* dl, sd, hh
Mapleleaf begonia *(Begonia dregei)* wtp, fs-dl, m
Mayflower *(Hepatica triloba)* div, w, dl-st, SD
Oncidium pulcheleum *(Oncidium pulcheleum)* wtp, sd, hh, dl
Oxalis *(Oxalis rubra)* bdv, m, fs

Parlor palm, miniature *(Chamaedorea elegans bella)* dl, sd, as
Partridgeberry *(Mitchella repens)* m, dl, as
Pilea depressa *(Pilea depressa)* wtp, dl, nwh
Pipsissewa *(Chimaphila umbellata)* m, dl-st, as
Rainbow flower (*Achimenes* hybrids) wtp, fs-dl, n, hh, drp
Rattle-snake *(Crassula teres)* sd, fs
Silver orchid *(Erythrodes noblis argyrocentrum)* wtp, sd, dl,as
Spider Plant (*Chlorophytum comosum* "vittatum") fs-dl, m
Spring Beauty *(Claytonia virginica)* m, fs-st
Star-of-Bethlehem *(Campanula isophylla alba)* ctp, fs-dl
Strawberry geranium *(Saxifraga sarmentosa)* ctp, fs, sd, hh
Sweet flag, dwarf *(Acorus gramineus pusillus)* w, fs
Tillandsia iolantha *(Tillandsia ionantha)* wtp, dl, sd, hh
Violet, dog *(Viola conspersa)* ctp, dl-st, m
Wintergreen, spotted *(Chimaphila maculata)* w, dl, as
Wood sorrel (*Oxalis* spp.) m, dl, as